Engineering GCSE

Engineering GCSE

Secon

Mike To

*ex-Vice Pr
Brookland*

ELSEVIER

AMSTERDAM • BOSTON • HEIDELBERG • LONDON • NEW YORK • OXFORD
PARIS • SAN DIEGO • SAN FRANCISCO • SINGAPORE • SYDNEY • TOKYO
Newnes is an imprint of Elsevier

Newnes

Newnes
An imprint of Elsevier
Linacre House, Jordan Hill, Oxford OX2 8DP
30 Corporate Drive, Burlington, MA 01803

First published 2002
Reprinted 2003, 2004
Second edition 2005

British Library Cataloguing in Publication Data
A catalogue record for this book is available from the British Library

Library of Congress Cataloguing in Publication Data
A catalogue record for this book is available from the Library of Congress

ISBN 0 7506 6576 9

For information on all Newnes publications visit
our website at www/newnespress/com

Typeset by Charon Tec Pvt. Ltd, Chennai, India
Printed and bound in Great Britain

Working together to grow
libraries in developing countries

www.elsevier.com | www.bookaid.org | www.sabre.org

ELSEVIER   BOOK AID
            International   Sabre Foundation

Contents

Preface   vii

Chapter 1   Design and graphical communication   1
   Unit 1 of the GCSE syllabus

Chapter 2   Engineered products   97
   Unit 2 of the GCSE syllabus

Chapter 3   Application of technology   227
   Unit 3 of the GCSE syllabus

Chapter 4   Maths and science for engineering   273
   Additional chapter for all students requiring
   further science and math's underpinning
   knowledge

Answers to numerical Test your Knowledge   321
   questions

Appendix: Engineering data   327

'How to' Index   333
Index   334

Preface

Welcome to the challenging and exciting world of engineering! This book is designed to help you succeed on a course leading to the vocational GCSE qualification in Engineering. It contains all of the essential underpinning knowledge required of a student who has never studied engineering before and who wishes to explore the subject for the first time.

About you

Have you got what it takes to be an engineer? The Engineering GCSE will help you find out and still keep your options open. Engineering is an immensely diverse field but, to put it simply, engineering, in whatever area that you choose, is about thinking *and* doing. The 'thinking' that an engineer does is both logical and systematic. The 'doing' that an engineer does can be anything from building a bridge to testing a space vehicle. In either case, the essential engineering skills are the same.

You do not need to have studied engineering before starting the GCSE course. All that is required to successfully complete the course is an enquiring mind, an interest in engineering, and the ability to explore new ideas in a systematic way. You also need to be able to express your ideas and communicate these in a clear and logical way to other people.

As you study this course, you will be learning in both a classroom and a workshop environment. This will help you to practice the things that you learn in a formal class situation. You will also discover that engineering is fun – it's not just about learning a whole lot of meaningless facts and figures!

About Engineering GCSE

The GCSE Engineering is a nationally recognised and accredited qualification designed to provide you with a choice of routes into further education or employment. The GCSE in Engineering is a double award equal to two GCSEs. It is therefore twice the size of most GCSEs and represents twice the work that would normally be associated with a single award. The GCSE in Engineering is a 'vocational GCSE'. This means that it is directly relevant to anyone who may be thinking about a career in engineering.

The first three chapters that make up the GCSE course cover Unit 1: *Design and Graphical Communication*, Unit 2: *Engineered Products*, and Unit 3: *Application of Technology*. In the first two chapters you will build up a portfolio of your work. In Chapter 1 (Design and Graphical Communication) we have called this your 'Personal Design Portfolio' and you should take particular pride in its development because it says a great deal about you. You will continue to develop your portfolio in Chapter 2 (Engineered Products) but in this chapter you will also have the opportunity to produce an engineered product. Your teacher or lecturer will suggest what this should be and he or she will also help to ensure that you have the right materials and resources to complete the task!

In Chapter 3 (Application of Technology) you will investigate the use of technology through a series of open-ended case studies. It's also worth remembering that a study of engineering requires a good grasp of science and mathematics. So, if you intend to study engineering at a higher level, it is important that you develop your skills in these subjects, at the same time as studying for your GCSE in Engineering.

How to use this book

This book covers the three units that make up the Engineering GCSE double award programme. The first three chapters are devoted to each of the three assessed units and each of these chapters contains text, key points, 'test your knowledge' questions, activities, and review questions.

Since many GCSE students have difficulty with science and mathematics, we have included an extra chapter on *Maths and Science for Engineering*. Whilst this additional chapter is not part of the GCSE double award and is therefore not part of the GCSE assessment, it has been designed to complement the other chapters and also to provide the essential maths and science underpinning knowledge that will allow students to progress to engineering courses at a higher level. Furthermore, unlike GCSE Maths and Science courses, all of the topics, questions and examples in this fourth chapter relate directly to engineering applications.

The second edition of *Engineering GCSE* has been considerably enhanced and expanded to include a number of new topics. These include new activities on interpreting client design briefs and using CAD packages, and new sections on quality control and quality assurance, health and safety, casting and milling, and systems and control technology.

The 'test your knowledge' questions are interspersed with the text throughout the book. These questions allow you to check your understanding of the preceding text. They also provide you with an opportunity to reflect on what you have learned and consolidate this in manageable chunks.

Most 'test you knowledge' questions can be answered in only a few minutes and the necessary information can be gleaned from the surrounding text. Activities, on the other hand, require a significantly greater amount of time to complete. Furthermore, they often require additional library or resource area research coupled with access to computing and other information technology resources.

Activities are the means by which you generate the portfolio evidence needed to satisfy the assessment requirements for Units 1 and 2 (note that Unit 3 is assessed by means of a written examination). As you work through this book, you will undertake a programme of activities as directed by your teacher or lecturer. Don't expect to complete *all* of the activities in this book – your teacher or lecturer is expected to ensure that those activities that you do undertake will generate ample assessment evidence. Activities also make excellent vehicles for gathering the evidence that can be used to demonstrate that you are competent in core skills.

These essential features of *Engineering GCSE 2e* provide vital practice material for the GCSE course, but are not intended as an assessment instrument to replace Edexcel assessment materials.

Finally, here are a few general points worth noting:

- Allow regular time for reading – get into the habit of setting aside an hour, or two, at the weekend to take a second look at the topics that you have covered during the week.

- Make notes and file these away neatly for future reference – lists of facts, definitions and formulae are particularly useful for revision!

- Look out for the inter-relationship between subjects and units – you will find many ideas and a number of themes that crop up in different places and in different units. These can often help to reinforce your understanding.

- Don't be afraid to put your new ideas into practice. Remember that engineering is about thinking *and* doing – so get out there and *do* it!

- Lastly, I hope that you will find some useful support material at my website, www.key2study.com

Good luck with your Engineering GCSE studies!

Mike Tooley

Chapter 1 | Design and graphical communication

Summary

This chapter covers Unit 1 of the GCSE engineering curriculum. It will introduce you to the skills associated with designing an engineered product or service and the means by which you can effectively communicate your ideas to other people. You will first study and practice the skills associated with designing an *engineered product* or an *engineered service* and then you will learn how to communicate your ideas to other people.

As you study this chapter you will learn the following:

- What the design process is and how it works
- How to express a design problem
- How to present a design brief
- How to select from a range of alternative design solutions
- How to present your chosen design solution to a client
- How engineering drawings are used to present information
- How to use different techniques for communicating information.

During your study of the chapter you should build a *portfolio* of your work. This is a collection of notes, sketches, drawings, presentations and other work that you produce as you study the chapter. Your portfolio will provide you with a valuable collection of material that shows what you did and what you learned during the chapter.

1.1 Introduction

Being able to design something is a fundamental engineering skill that you will develop as you progress through this chapter. However, being able to design something is not the end of the story. Equally important is that you should be able to communicate your design to other people. Indeed, even the most basic design will be hopelessly flawed if you cannot explain to people what it is about and, equally important, how to make it!

As an engineer, you might be involved with the design of an engineered product that varies from something as basic as an adjustable spanner to something as complex and sophisticated as a military aircraft. This chapter will help you understand the design process and how to communicate your ideas to other people.

You will already be familiar with engineered products, such as cars, aircraft, radios, computers, DVD players, videos and hi-fi systems. Not all engineered products are as complex as these. Something as simple as a screwdriver, a corkscrew, a can opener or a light bulb is also an engineered product.

Engineered services include activities like motor vehicle maintenance, aircraft maintenance, electronic servicing and motorway maintenance. These services need to be designed and communicated in just the same way that applies to engineered products.

You should begin work on your *personal design portfolio* (PDP) as soon as you start the chapter and then continue to collect evidence of your work as you progress through the chapter. Your PDP will also help you to collect the evidence that you need to demonstrate that you have become proficient in key skills.

1.2
The design process

The design process is the name given to the various stages that we go through when we design something. Each stage in the process follows the one that goes before it and each stage is associated with a particular *phase* in a design project.

Before going much further, it is worth remembering that designing is not always about creating a brand new product or service. In fact, in most cases it is about improving or modifying an *existing* product. When designing a brand new product or service, the design process will have more stages because we usually need to consider a wider range of options and alternative solutions than when we are simply modifying or redesigning an existing product.

A typical design project involves the following tasks (see Figure 1.1):

- Understanding and describing the problem
- Developing a design brief with the client
- Carrying out research and investigation
- Generating ideas using techniques such as brainstorming and mind mapping
- Investigating solutions and applying scientific principles
- Developing an agreed set of design specifications
- Communicating the design solution using appropriate engineering drawings
- Realising the design solution
- Evaluating the design solution.

We shall look at what actually makes up each of these tasks later on. For now, you only need to be aware that designing something involves a series of tasks and each of these forms an important stage in reaching your eventual goal, an *engineered product* or an *engineered service*. If you find this difficult to remember, Figure 1.2 shows the design process described only in terms of the key words that relate to each of the individual stages.

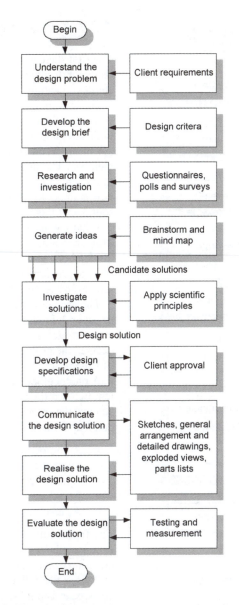

Figure 1.1 *The design process*

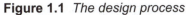

Activity 1.1

Think about an engineered product that you use in everyday life such as a bicycle, a hi-fi system or a DVD player. Write down a list of three or four features or improvements that you would like to incorporate in the product. For each feature or improvement suggest what modifications or changes would have to be made. Present your work in the form of a single A4 page and include hand-drawn sketches where appropriate. Do not forget to put your completed work in your PDP!

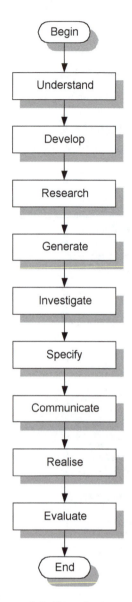

Figure 1.2 *The design process described in terms of the key words for each stage of the process*

1.2.1 The design problem

The first stage in the design process is that of identifying the problem that your design is to solve. Until you understand what the problem is about you cannot begin the next stage of the process.

For example, let us imagine that you have been asked to help the secretary of your school's computer club get a mail shot out to members with dates and times of meetings. The club secretary has given you an A4 information sheet and has asked you to copy and circulate it to members. He suggests that you make photocopies of the page, and then fold each copy before inserting it into individual envelopes that you will address to each member. By now, this might be beginning to sound as if this could be a time consuming task!

What you have not done is ask yourself what the problem really is. Think about it. In this case it is simply a matter of getting information to members. Posting letters to individual members is just one way of tackling the problem. Instead, let us think about the problem first and then examine different ways of solving it.

The problem *is not* about photocopying, folding and posting sheets of paper. It is actually about finding an effective way for the club secretary to get information about meetings to club members. Having defined the problem you can go on to the next stage, thinking of ways of solving it.

A photocopied mail shot is one way that you *could* solve the problem but is it the best way? Let us think of other ways that you could achieve the same goal. Here are just four possibilities that you might come up with:

- A photocopied mail shot sent to all members (the club secretary's original idea)
- A notice placed on a notice board
- An advertisement placed in the school newspaper
- A 'meetings' page placed on the club's Web site.

The first solution is going to involve you in a lot of work and some expense whilst the second is easy to do but is unlikely to gain the attention of all club members unless they are in the habit of regularly looking at the notice board. The third solution assumes that the school newspaper is published regularly and in advance of club meetings. It also presupposes that club members all read the school newspaper!

The last possibility is probably the most promising. Since computer club members are almost certain to have an Internet connection they are all likely to be regular visitors to the club's Web site. The next

Activity 1.2

In relation to Activity 1.1, select just one of the features or improvements that you have identified for your chosen engineered product. Now try to think of the 'problem' that has led you to suggest that this feature needs some attention. Summarise the problem in a single sentence and add this to your PDP.

stage of this particular 'design project' would be that of discussing the four potential solutions with the club secretary. Hopefully he will agree with your recommendation that the fourth solution is the easiest and the most effective!

1.2.2 The client

When a design engineer develops a design, he or she is usually working directly or indirectly for a *client*. Often the client may be an individual, a company or an organisation outside that for which the designer works. However, in some cases the client may be another department or division of the *same* company. In order to distinguish between these two situations, we sometimes refer to clients as being either *internal clients* or *external clients*. So, an *internal client* is an individual or department who is employed within the same company or organisation that you work for whilst an *external client* is someone that is employed by another company or organisation.

The relationship between a designer and his or her client is an important one. The client may only have a rough idea of what the end result of the design will look like and how it might work. In other cases the client may be able to *suggest* just how the product or service will work and what it will look like. In this case we would say that the client has a *preconceived* idea of what the product or service will be. Sometimes this is a good thing because it gives the designer a starting point to work from. In many cases, however, preconceived ideas only serve to limit the range and effectiveness of the solution that the designer comes up with.

1.2.3 The design brief

Before you can begin a design, you need to understand what your client needs. You do this by developing a *design brief*. A design brief is a statement that identifies what is needed to solve the problem. The design brief must not be vague nor should it be too lengthy. It must also be worded in such a way that it avoids any preconceived ideas of what the solution might be. The design brief should not actually suggest the solution to the problem. Instead, it is for you, the designer, to suggest ways of solving the problem thereby satisfying the design brief and solving the client's problem in the most effective and appropriate way.

How to prepare a design brief

✓ Start by checking that you fully understand the design problem.

✓ Summarise the design problem using short sentences or bullet points.

✓ Write down, without actually suggesting what the solution is, what you need to do to solve the design problem.

Test your knowledge 1.1

Decide on whether the client is internal or external in each of the following situations:

1. Your teacher asks you to design a lighting console that can be used in conjunction with a forthcoming school production.
2. Whilst you are out on work experience, the company that you are placed with asks you to assist with a design project that has been commissioned by a leading Formula 1 race team.
3. Your older brother asks you to help him design and build a trailer that will be used to transport a go-kart to race meetings.
4. A local engineering company asks your class to produce a series of design ideas for a new workshop that they have agreed to sponsor.

Test your knowledge 1.2

Explain what a preconceived idea is and why they should be avoided.

Key point

The wording of a design brief is important. It is a good idea to only use simple words and agree these with your client before you begin to look for ways of solving the problem. If the client provides you with the design brief (rather than you having to develop it for the client) you need to make sure that *you* understand it fully before going any further!

In order to make your design brief more useful you can

✓ Use only simple words.

✓ Avoid re-stating the design problem!

1.2.4 Research and investigation

Having developed the design brief with your client, the next stage in the design process is that of carrying out some initial research and investigation. In some cases, your client may have already carried this research out before asking you to develop the design brief. In other cases, you may find that you need to carry out the research on behalf of your client.

Depending upon the design brief, the research that you carry out might include:

- Assessing the likely market potential for the product or service
- Evaluating competing or existing products that meet a similar need
- The views of potential end-users.

Research is usually carried out by one or more of the following methods:

- Questionnaires
- Polls
- Surveys.

Questionnaires consist of a series of questions that are given to a particular *target group* to answer. Sometimes, the target group is taken as a representative sample of the population but sometimes you may decide to restrict a questionnaire to a particular set of people (e.g. those over 25, those with an annual income of more than £25,000, or those who own a personal computer (PC)). Every member of the target group that completes a questionnaire is known as a *respondent*.

Questionnaires must be clear, relatively simple and understandable to all and should consist of mainly 'yes/no' or single-word answers. You should avoid making a questionnaire too long as people will not want to spend too much time on it! It is also worth noting that it can take *you* some time to carry out an analysis of the results of a detailed questionnaire.

Polls are usually a lot simpler than questionnaires and usually involve respondents making a choice from a restricted selection. For example, 'Which of these four colours do you prefer?', 'Which of these three styles do you like the most?' and so on.

Polls are usually easy to carry out and may be carried out with groups of people (using, say, a simple show of hands) as well as with individuals. The results of a poll can usually be obtained very quickly and often will not require much in the way of further analysis or interpretation.

Test your knowledge 1.3

Arrange each of the following under one of two headings:
'1. A design brief should ...' and '2. A design brief should not ...'

(a) be kept as short and simple as possible
(b) use straightforward words
(c) suggest the solution to the problem
(d) describe what the problem is about
(e) be agreed with the client
(f) suggest a variety of different solutions
(g) include detailed drawings and specifications.

Test your knowledge 1.4

Consider the following statements and comment on their use as a design brief for the civil engineering company that subsequently went on to design and build the Channel Tunnel:

(a) To build a tunnel between Dover and Calais.
(b) To provide a rail link between England and France.
(c) To provide a fast and convenient alternative to conventional cross-channel car ferry services.
(d) To significantly reduce channel crossing times and to make it possible to travel to the continent by train quickly and easily.
(e) To make it possible to travel from London to Paris by road or rail in less than 4 h.

Which of these statements is the best design brief and why?

Surveys combine some of the elements of both questionnaires and polls. Surveys are normally carried out with respondents on a 1:1 basis. The person carrying out the survey will usually explain each of the questions to the respondent and then note down the response. Surveys usually take some time to carry out and the results can take some time to analyse. To put all of this into context consider the following example.

Quick Byte, a chain of computer shops, intends to open a dropin PC Clinic in each of its shops. In order to assess the likely demand for this new service, the company decides to carry out some market research using a questionnaire that will be given out in shops to each customer.

In order to provide people with an *incentive* for completing the questionnaire, Quick Byte will hold a draw with prizes that respondents will be entered into when they return the questionnaire. The Quick Byte questionnaire is shown in Figure 1.3.

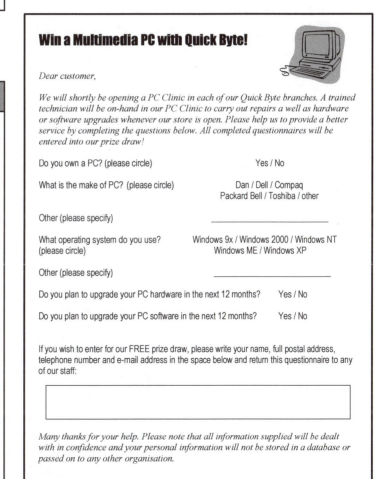

Figure 1.3 *The Quick Byte survey questionnaire*

Test your knowledge 1.5

Classify each of the following research activities as either a survey, a poll or a questionnaire:

(a) At a class meeting, your teacher asks the class to vote on which of three class members should become class representative on the school's Student Council.

(b) A Local Councillor visits you at home to find out your views on the effectiveness of a recent Road Safety campaign.

(c) Your local Youth Club asks members to vote for the winner of its Rising Star talent competition.

(d) You receive a letter from your local hi-fi dealer asking you to complete a form that requires you to make some simple 'yes/no' responses relating to its after-sales service.

The results of Quick Byte's questionnaire are as follows:

Do you own a PC?	2306 Yes, 391 No
What is the make of PC?	258 Compaq, 421 Dan, 381 Dell, 595 Packard Bell, 212 Toshiba, 395 other
What operating system do you use?	821 Windows 9x, 231 Windows 2000, 37 Windows NT, 209 Windows ME, 871 Windows XP
Do you plan to upgrade your PC hardware in the next 12 months?	871 Yes, 1301 No
Do you plan to upgrade your PC software in the next 12 months?	2031 Yes, 259 No

The results of Quick Byte's survey can be best illustrated using a series of charts rather than simply quoting a series of figures. Furthermore, since Quick Byte is interested in the *proportion* of customers that respond to several of these questions (i.e. how many reply 'Yes' compared with how many reply 'No') rather than the exact number, a good way to illustrate the data is by using a series of pie charts (see Figure 1.4). Quick Byte will use bar charts to illustrate the response to the remaining questions (i.e. those that relate to make of PC and the operating system that it uses). When constructing a *bar chart*, the height of the bar corresponds to the proportion of responses. In practice, we would choose a convenient scale to draw the bar chart, such as 20 mm = 1000 responses.

When constructing a *pie chart*, the angle of the segment (in degrees) corresponds to the proportion of responses. Since the complete pie chart occupies an angle of 360° the angle of the segment that you need to draw can be calculated from the following formula:

$$\text{Segment angle} = (\text{Number of responses/Total responses}) \times 360°$$

How to write a questionnaire

✓ Think about what information you need to get from the questionnaire and why you need it.

✓ Give the questionnaire a title and write a brief introduction which will explain to respondents why you need the information and what you intend to do with it. Do not forget to tell them why the questionnaire will benefit them. Also decide whether you need to offer some incentive for completing the questionnaire.

✓ Number your questions and write each one down using simple words. Make sure that there is a limited range of answers to each question. Use tick boxes wherever possible.

The following data was obtained from a questionnaire distributed by a lawn mower manufacturer:

1. What type of lawn mower do you own?
   (a) Electrically powered cylinder type: 757
   (b) Electrically powered rotary type: 931
   (c) Petrol engine powered cylinder type: 231
   (d) Petrol engine ride-on type: 102
   (e) Other (e.g. solar powered or battery operated): 39

2. How do you rate your lawn mower in terms of ease of use?
   (a) Very easy to use: 379
   (b) Easy to use: 697
   (c) OK to use: 553
   (d) Difficult to use: 394
   (e) Very difficult to use: 37

3. How do you rate your lawn mower in terms of the quality of its cut?
   (a) Excellent: 220
   (b) Good: 781
   (c) Fair: 599
   (d) Poor: 245
   (e) Very poor: 209

4. Are you thinking about changing your lawn mower in the next 18 months?
   (a) Yes: 494
   (b) No: 991
   (c) Don't know: 505

Illustrate the results of this survey using bar charts for questions 1 to 3 and a pie chart for question 4. Do not forget to add this work to your PDP!

✓ Do not forget to thank respondents for completing the questionnaire!

In order to make your questionnaire more useful you can

✓ Invite respondents to give further information where appropriate. Use wording such as: 'other – please specify …'

Do you own a PC?

What is the make of your PC?

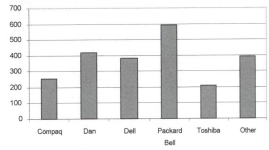

What operating system do you use?

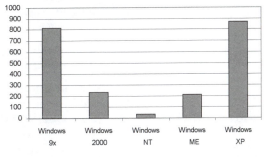

Do you plan to upgrade your PC hardware in the next 12 months?

Do you plan to upgrade your PC software in the next 12 months?

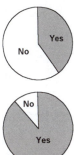

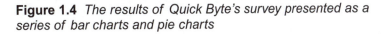

Figure 1.4 *The results of Quick Byte's survey presented as a series of bar charts and pie charts*

How to draw a bar chart

✓   You will need an A4 sheet of squared paper and a ruler. You will also need a calculator and an HB pencil or a drawing pen.

✓   Decide on how large you will make the bar chart and then draw the base line for the chart.

✓   Decide on the scale that you will use (e.g. 1 cm = 1000 responses) and then work out the size of the first bar by calculating its height. Do this by taking the number of responses that correspond to the first bar and multiplying by the scale factor (e.g. 1/1000 cm). Mark this height and then complete the rectangle for the first bar. Write the name of the bar below or inside the bar.

✓   Repeat for all other bars.

✓   Write a title below your bar chart.

In order to make your bar chart more useful you can

✓   Shade the bars using different colours.

✓   Write the scale that you used in brackets below the title.

How to draw a pie chart

✓   You will need an A4 sheet of plain paper, a compass and a protractor. You will also need a calculator and an HB pencil or a drawing pen.

✓   Decide on how large you will make the pie chart and then set the compass to draw a circle of this size.

✓   Work out the size of the first segment by calculating its angle. Do this by taking the number of responses that correspond to the first segment and dividing this by the total number of responses. Multiply the result by 360 to find the number of degrees, then use your protractor (placed on the centre of the circle) to mark the angle. Write the name of the segment inside the segment.

✓   Repeat for all other segments (you will not need to calculate the angle for the last segment because it will simply be what is left over when all the other segments have been drawn!).

✓   Write a title below your pie chart.

In order to make your pie chart more useful you can

✓   Shade the segments of the pie chart using different colours.

✓   Write the total number of responses in brackets below the title.

Key point

Polls and questionnaires can be used to assess the likely market potential for a product or service. They also prove a way of evaluating existing or competing products.

Activity 1.3

Devise a questionnaire that can be sent to members of your class in order to find out whether they have a PC at home and what they use it for. Ask no more than 10 questions. Distribute your questionnaire to at least 10 people, analyse the results of the questionnaire and present your analysis in the form of a series of bar charts or pie charts. Do not forget to add this work to your PDP.

1.2.5 Generating ideas

You may often find that you generate ideas without too much effort. Just thinking about a problem is likely to lead you to one or more ideas that may help to solve it. Imagine yourself on a cross-country run standing on the bank of a river. You need to get to the other side of the river but there is no bridge in sight. You do not know how far the next bridge is nor in which direction you should run to reach the nearest bridge. You also do not know how deep the water is. What do you do? Your first idea might be to see if there is anyone around to ask where the nearest bridge is or how deep the water is. Your next idea might be to find a stick to probe the water and find out how deep the water is. After that, you might think about finding someone with a boat that can ferry you to the other side. Having collected together a few ideas you will then have some options from which to choose. Which of these you actually decide to follow might depend on how anxious you are to complete the course and your assessment of whether or not the option that you have chosen is likely to bear fruit!

In everyday life, ideas often seem to flow naturally. When designing an engineered product or service this is not always the case. Furthermore, if you can only come up with a limited number of ideas (say one or two) you might need to generate more ideas to provide you with a wider range of alternatives or *options*.

In order to generate ideas we can make use of one or two tried and tested ideas. The first of these is called *brainstorming*.

In brainstorming a group of people sit around and fire ideas at one another. There are several basic rules for brainstorming:

- Everyone in the group must contribute and has an equal right to be heard.

- All ideas (however unlikely or preposterous) must be treated with equal respect.

- Everything should be written down so that no ideas are lost (usually one member of the group is made responsible for this and ideas are recorded on a flip chart so that all can see what has been written down).

- Adequate time should be set aside for the exercise and there should be no interruptions.

- It is important to avoid probing ideas too deeply. This can be left until a later stage.

- Agree, at the end of the session a selection (typically three or four) of ideas that should be considered as *candidates* for carrying forward to the next stage of the process. These are the ideas that the group considers (by poll, if necessary) to be the most feasible in terms of satisfying the design brief. Do not, at this stage, reject the other ideas – you might need to come back to these later!

At first sight, some ideas may be considered less credible or less serious than others by some of the members of the group (we often describe such ideas as being *off-the-wall*). Nobody in the group should be made to feel bad or inferior if other members of the group consider their ideas strange or unworkable. Some of the most innovative engineering projects have resulted from brainstorming sessions that have unearthed ideas that, at first sight, have been considered unworkable by the majority of those involved!

Another technique that is used to generate ideas is *mind mapping*. A mind map is a sketch or drawing that allows you to identify all the factors that need to be taken into account when developing a solution to a design brief. The name of the product or service appears at the centre of the mind map and each of the solutions and other factors are

How to conduct a brainstorming session

✓ You will need a flip chart and some markers.

✓ Form a small group to carry out the brainstorming (four to six people is ideal).

✓ Set aside sufficient time to meet and choose a venue for the meeting where you will not be disturbed.

✓ Begin by explaining the design problem and make sure that everyone understands it. It is a good idea to write down the design brief on the first page of the flip chart then tear this page off and pin it up so that everyone can see it.

✓ Decide on who will write down the ideas that you generate using the flip chart (this person can also contribute to the discussion).

✓ Invite everyone to contribute to the discussion and make sure that they know that all of the ideas generated will be equally valued by the group. Make sure that everyone understands this rule and that they all know the reason for it!

✓ At the end of the session, summarise the ideas that have been put forward and thank everyone for their contributions.

In order to make your brainstorming session more useful you can

✓ Summarise the ideas from the flip chart using a single sheet of A4 paper and circulate this to all those who took part.

✓ Set a deadline for comments and any further suggestions to be sent to you.

Activity 1.4

Carry out a brainstorming activity in groups of four or five. One person should be given the task of recording all of the ideas put forward using a flip chart. Agree a fixed time for the brainstorming session (say 35 min). At the end of this time the group should select three solutions to investigate further. The brainstorming activity should be devoted to finding ways of solving the following design problem and design brief:

The design problem: A high proportion of road accidents that occur at night are due to drivers falling asleep at the wheel.

The design brief: Develop a means of significantly reducing the number of road accidents that occur as a result of driver fatigue.

Write a brief account of your brainstorming session and summarise each of the solutions that were put forward. Do not forget to add this to your PDP.

placed around it. The map can then be progressively expanded as more detail is added.

There are a number of advantages of using a mind map to generate ideas and to understand the relationship between them. These include:

- The design brief (or design problem) appears in the centre of the map and it is thus very easy to see how all of the potential solutions and any other factors relate to it.
- The links that exist between solutions and other factors can be immediately recognised.
- The map can be easily grasped without having to read a lot of words.
- It is easy to extend a map or add more information to it.
- A mind map can help to stimulate thought and aid understanding.
- It is often easier and faster to create a mind map than spend a lot of time putting your ideas in writing!

To explain how a mind map works let us assume that the Head of Technology in your school has been asked to find some way of taking a series of aerial photographs of the school's outdoor swimming pool using a compact digital camera. The Technology Department has been given a limited sum towards the cost of materials and there appears to be two ways in which the problem could be solved, using some form of platform or some form of aircraft. The platform solution could be based on a number of options, namely a scaffolding tower, a pneumatic mast or some form of extending arm. The aircraft solution could use a piloted aircraft (an expensive option), a radio-controlled model aircraft or a tethered balloon. The piloted aircraft

Key point

A mind map is a sketch that allows you to identify all the factors that need to be taken into account when developing a solution to a design brief. The name of the product or service appears at the centre of the mind map and each of the solutions and other factors are placed around it. The map can be expanded with more detail as required.

could be a plane, a helicopter or a hot-air balloon whilst the radio-controlled aircraft could be a plane or a helicopter. These solutions are illustrated in the mind map shown in Figure 1.5.

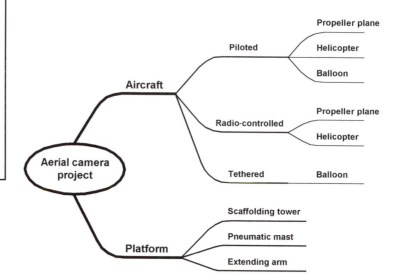

Figure 1.5 *Mind map used to identify solutions to the problem of taking aerial photographs of the school's outdoor swimming pool*

How to construct a mind map

✓   You will need a large sheet of paper (ideally A3 or larger) and some coloured felt-tip pens or markers.

✓   Start by writing down a short phrase that describes the design brief or states the design problem. This should be placed in an oval shape in the centre of the sheet.

✓   Add thick lines radiating from the oval (as many as are necessary) corresponding to each main solution that you have identified.

✓   Print the name of each solution above the line.

✓   Keep on extending the map (moving outwards from the centre) showing further details.

In order to make your mind map more useful you can

✓   Use different colours to distinguish the lines leading to each main solution.

✓   Use pictures, arrows and notes to provide more information.

✓   Make the map look attractive and clear. That way, people will remember it!

A radio telescope is to be built in the crater of an extinct volcano in a mountainous region of an island in the Pacific Ocean. The construction materials are to be transported from Westport to the construction site that is approximately 70 km from the port. The materials are large and expensive and easily damaged by mechanical shock, vibration, rough handling and high temperatures. Figure 1.6 shows the main geographical features of the island together with transport links.

The following additional information has been provided by a local agent:

- The bridge at Fisher Gully has collapsed.
- The Central Desert region is rough and extremely hot.
- The river is impassable at Indian Rapids.
- The only aircraft available is a twin-engined transport plane.
- Mules are only capable of carrying light loads over rough terrain.
- There are several lorries available and one diesel locomotive with rolling stock at South Bay.

A first attempt at developing a mind map showing the various options for transporting the construction materials is shown in Figure 1.7. Complete the mind map and use to identify a preferred solution. Present your work in the form of a folder that can be included in your PDP.

Figure 1.6 *See Test your knowledge 1.8*

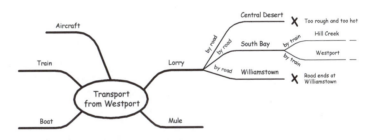

Figure 1.7 *See Test your knowledge 1.8*

Create a mind map that illustrates the range of possible solutions and other factors in relation to Activity 1.4. Draw your mind map on a sheet of A3 paper and use colours and sketches to add impact and interest. When you have completed this work do not forget to include it in your PDP.

1.2.6 Investigating solutions

Having generated a number of ideas – whether by using brainstorming, mind mapping or by some other technique – you will need to narrow down your potential solutions to those that are most likely to provide you with the optimum solution. Each of those that you select for further investigation can be referred to as *candidate solutions*.

Each candidate solution should be investigated and the successful candidate solution can then be identified by evaluating each against a set of *criteria*, such as physical properties, (e.g. size and weight), cost of materials, ease of manufacture, reliability and so on.

When selecting the final *design solution* from the set of candidate solutions you need to ask yourself a number of questions, including:

- What does the product or service do? Does the proposed solution solve the problem described in the brief?

- What materials or resources are required to manufacture the proposed solution or provide the proposed service? What tools, equipment and people are required?

- What must the product or service look like? How easily can this be achieved using the proposed solution? Are there any other *aesthetic* factors that need to be taken into account?

- What physical properties (i.e. size, weight, strength, etc.) are required? Can these requirements be met using the proposed solution? Does the solution meet all the physical requirements?

- Are there any problems or particular considerations that need to be borne in mind in relation to Health and Safety? Are there any environmental issues that need to be considered?

- How easy will it be for the client, customer, or *end-user*, to use the proposed solution? Are there any *ergonomic* factors that need to be taken into account?

The term *ergonomics* comes from two Greek words 'ergo' meaning 'a task' and 'nomos' meaning 'a law or a rule'. Ergonomics (sometimes also known as *human engineering*) is the study of how the human body relates to its environment. When considering design solutions we need to give some thought as to how human beings (i.e. end-users) will interact with what it is that we are designing!

One good way of comparing a set of candidate solutions is the use of an *evaluation matrix*. This is simply a table that lists the design criteria (in rows) and each of the candidate solutions (in columns). When you construct an evaluation matrix you put a mark or a score in the cells corresponding to your assessment of each candidate solution against each criterion.

How to use an evaluation matrix

✓ Make a list of the candidate solutions that you have identified from your brainstorming or mind mapping sessions.

✓ Write down a set of design criteria against which you will use to evaluate each solution.

✓ Create the matrix using columns for each solution and rows for each criterion.

Test your knowledge 1.9

Explain what is meant by:

(a) a candidate solution
(b) aesthetics
(c) ergonomics.

Test your knowledge 1.10

Figure 1.8 shows a partially complete evaluation matrix for three candidate solutions for containers that will be used to supply orange juice to consumers. When you have completed this work include it in your PDP.

Key point

Design criteria are a list of characteristics that we require the final design solution to satisfy. Some of these may be considered 'essential' whilst others may just be 'desirable'. In order to reduce the number of candidate solutions to a final design solution we need to check to see whether the design criteria are met.

✓ Decide on which of the criteria are *essential* and which are *non-essential*. Mark them accordingly.

✓ Evaluate each solution by placing ticks and crosses (or question marks if you are uncertain) in each of the cells.

✓ Count the ticks (or total the scores) and decide which solution scores the highest.

✓ Repeat this process for each candidate solution.

✓ Eliminate any solution that fails to satisfy a criterion that you have rated as *essential*.

In order to make your evaluation matrix more useful you can:

✓ Award a points score instead of using ticks and crosses. A typical scoring system might be based on scores of 0 to 4 where 0 indicates that the solution completely fails to satisfy the criterion and 4 indicates that the solution fully meets the criterion.

✓ Rate your solutions as 'best', 'alternative', 'possible', etc.

Design criterion	Candidate solution			
	Plastic bottle with screw top	Glass bottle with foil seal	Plastic carton with screw top	Waxed paper box with foil seal
Must be low-cost		✓		
Must be hygienic (essential)		✓		
Must be recyclable		✓		
Must be resealable (essential)		✗		

Figure 1.8 *See Test your knowledge 1.10*

1.2.7 Applying scientific principles

As you work towards your design solution you will need to apply scientific principles in order to be sure that your idea will work as planned. The scientific principles that you need to apply will depend on your design project but typically might involve the relationship between forces, loads, areas, volumes, speed, acceleration, energy and power.

As scientific principles are so important, we often say that science *underpins* our understanding of engineering. In a design project it is this understanding that gives us the confidence that our design will actually work!

As an example, let us assume that our design brief involves moving an object that has a mass of 100 kg through a vertical height of 10 m. To meet our client's requirements we have to perform this operation in under 40 s. We intend to use an electric motor to power the hoisting mechanism that has an efficiency of 40% (in other words, 40% of the electrical power input is converted into useful mechanical energy).

The question remains as to whether this arrangement will actually satisfy the essential design criteria of being able to lift the load in under 40 s!

In this sort of situation it is always a good idea to use a sketch in order to picture what is going on. Figure 1.9 shows the arrangement of the hoist labelled with the various things that we know. We have also introduced some symbols (m, g, h, etc.) to represent the quantities.

The scientific principles that we need to consider are as follows:

- The energy required to raise the mass is proportional to the product of its mass, the gravitational force, and its height (thus, energy $= m \times g \times h$).
- Power is the rate of using energy (thus, power $=$ energy/time).
- Efficiency is the ratio of output to input usually expressed as a percentage (thus efficiency $=$ (output/input) $\times$ 100%).

The energy required to lift the load $= m \times g \times h = 100\,\text{kg} \times 9.81\,\text{m/s}^2 \times 10\,\text{m} = 9.81\,\text{kJ}$. The rated motor input power $= 1\,\text{kW}$ and the mechanical output power $= 40\% \times 1\,\text{kW} = 400\,\text{W}$.

Since power $=$ energy/time we can infer that time $=$ energy/power. Thus the time taken to lift the load can be calculated from: Time $= 9.81\,\text{kJ}/400\,\text{W} = 9810/400 = 24.53\,\text{s}$.

The result (approximately 25 s) indicates that our design solution should easily be able to satisfy the essential design criterion of moving the load in under 40 s!

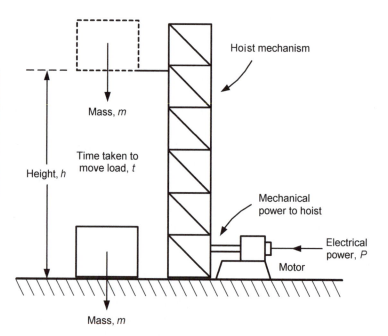

Figure 1.9 *Hoist arrangement*

Test your knowledge 1.11

Match each of the scientific principles in List 1 with the applications for design solutions given in List 2.

List 1

1. The work done moving an object is the product of the force applied and the distance from the point of application.
2. The amount of electrical charge transferred is the product of the current flowing and the time for which it flows.
3. The strength of a magnetic field around a wire is proportional to the current flowing in it.
4. When an electrical conductor is suspended in a magnetic field the force acting on it will be proportional to the current flowing.
5. The energy stored in a fly-wheel is proportional to the mass of the flywheel and its radius.

List 2

(a) A tape recorder.
(b) A push-and-go toy.
(c) A battery charger.
(d) A loudspeaker.

Key point

When carrying out a design project you should apply scientific principles to make sure that your idea will work as planned.

Activity 1.6

Investigate each of the following and identify the scientific principles that underpin them:

(a) A bicycle brake
(b) A glider
(c) A transformer
(d) A TV remote control.

Present your work in the form of a single A4 'fact sheet' for each device and briefly explain the underpinning scientific principles. Use sketches and diagrams wherever appropriate (there is no need to give any formulae). Make sure that your four completed 'fact sheets' appear in your PDP.

1.2.8 Design specifications

The following design brief is different from any that we have looked at so far. Can you say what is different about it?

Devise a system that will raise a 500 kg load measuring 1 m × 1 m × 1 m through a vertical height of 2 m and transport it across a workshop at a speed of 0.1 m/s over a distance of 10 m and then lower it to the ground safely.

I hope that you have spotted that this design brief is *very* precise! In fact, it's so precise that it quotes *figures* for the dimensions of the load, its mass, the height that it must be raised to, the distance that it must travel, and the speed that it must travel at. These figures are part of the *specification* for what the engineering system must do.

A specification is a detailed requirement that must be satisfied by the design solution. A specification must be written down. It must be accurate and precise. When we use a specification to help us design something we refer to it as a *design specification*.

A design specification helps the designer by giving him or her a very clear indication of the required performance of the product or service that is being designed. It is important to establish the design specification with your client at a very early stage in any design project.

The design specification is something that can be measured. The performance of your chosen design solution will need to be measured and compared with the design specification. A successful design solution is one that fully meets or conforms with the design specification.

Note that it may well be necessary to go back to your client if, for some reason, you find that your design solution only partially satisfies the design specification. Your client will have a view as to whether this is a problem that can be tolerated or whether it must be resolved.

British Standard (BS) PD 6112 identifies the items that should be included in a design specification. These *primary design needs* are as follows.

Title
The name or brief description of the product or service that you are designing.

History and background information
A general description of what the design problem is including, where appropriate, details of who the intended user is.

Scope of the specification
A description of each of the required features of the product or service. These may include characteristics such as size ('must be compact', 'must be at least 2 m wide', etc.), surface finish ('must be smooth', 'must be water resistant', etc.), and performance ('must be fast', 'must be powerful', 'must be efficient', etc.). The scope of the specification may also include any special needs of the people using the product, for instance if the users are elderly or have specific disabilities.

Definitions
Definitions should be included for any terms that your client may be unfamiliar with or that may have a special meaning. For example, a 'fuel strainer' may be defined as 'a fine gauze mesh that prevents dirt particles entering a carburettor'.

Conditions of use
You will need to give an indication of how, when and in what situation the product or service will be used. For example, 'continuous use', 'light domestic applications', 'commercial and industrial use', 'intermittent use', etc.

Characteristics
A simple break down of the different parts of the product, what their function is and any other features that the client needs to be aware of.

Reliability
An estimate of the working life of the product (e.g. 4000 h, 10,000 operations, etc.).

Servicing/maintenance features
A description of the maintenance requirements and the provision for repair or replacement of materials and parts that might wear out (e.g. coolant, lubricating oil, etc.) or repair in the event that the product fails or breaks down.

Other requirements that need to feature in a specification can include the following.

Ergonomics
Details about how users will interact with the product or service.

Aesthetics
Information relating to the appearance (e.g. shape, colour, surface finish) of the product or service.

Test your knowledge 1.12

Which of the following is NOT classified as a 'primary design need' within a design specification?

(a) A description of the design problem.
(b) A description of the client or end-user of the product or service.
(c) A description of each of the main features of the product or service.
(d) A description of the circumstances in which the product or service is to be used.
(e) A description of the tests and measurements carried out to ensure that the product or service conforms to the design specification.

Key point

A design specification is a complete summary of the required performance of the product or service that is being designed. A successful design solution is one that complies fully with the design specification.

Safety
Any specific safety features that relate to the product or its use.

Economics
The cost of producing, marketing and supplying the product or service when compared with the income that is generated from its sale or use.

Manufacture
Information relating to how the product will be made or assembled (e.g. 'using prefabricated parts assembled on-site', 'using fully automated assembly techniques', 'supplied in knock-down form for self-assembly', etc.).

Activity 1.7

Consider the following design problem, design brief and design criteria:

The design problem:
Many amateur astronomers use hand-held binoculars for studying objects in the night sky. Unfortunately, it is difficult to hold a pair of binoculars steady for a long enough period to make an adequate observation.

The design brief:
Develop a means of holding a pair of binoculars steady whilst an observation is being carried out.

The design criteria:
- Must be adjustable in azimuth and elevation
- Must be easily carried
- Must be easy to set up and dismantle
- Must be waterproof
- Must be robust and durable.

Produce a full design specification for this product using the headings given in the text. You should refer to a technical dictionary if you are unsure of the meaning of any of the terms (such as 'azimuth' and 'elevation'). Present your work as a folder and include it in your PDP.

1.3 Engineering drawing

Engineers rely heavily upon sketching and drawing as a means of communication. As an engineer you must be able to read and use working drawings as well as produce a selection of presentation drawings using both hand drawn and computer techniques. To avoid confusion, your engineering drawings must comply with recommended standards and conventions. You will also need to be able to read electrical/electronic, pneumatic/hydraulic and mechanical engineering drawings and identify a selection of the most commonly used symbols.

Activity 1.8

In this activity you are required to interpret a detailed design brief supplied by a client. The design brief is for a 17 tonne tipper lorry.

Start by reading the client's design brief (below) and then carry out further research in order to answer the questions. Make sure that you fully understand the design brief before attempting to answer the questions.

Design brief:

Engine	Diesel engine with a minimum of 180 BHP.
Gearbox	The gearbox shall have not less than six forward speeds.
Steering	The steering shall be power assisted.
Tyres	Tyres must comply with the tyre manufacturer's recommended specification for the gross vehicle weight. A spare wheel is to be fitted on a suitable carrier frame.
Suspension	Both front and rear suspension should comprise multi-leaf springs with shock absorbers.
Wheelbase	The wheelbase shall be not less than 3.9 m.
Tipper gear	The tipping gear is to be single ram front mounted. The angle of tip is to be sufficient to clear wet clay from the body with the vehicle stationary.
Cab	The cab shall be of the tilt type and should be fitted with a heater and a fire extinguisher. A reversing camera with horizontal and vertical vision shall be fitted. The cab shall be fitted with an audible warning device which can be operated from the rear of the vehicle. An in-cab audible overhead cable warning system shall be fitted.
Exhaust	The exhaust shall be a vertical type located between the cab and the vehicle body.
Warning indicators	Flashing trafficators and a roof mounted revolving amber hazard unit are to be fitted. An automatic reversing bleeper is to be provided.
Body	The body is to be of steel welded construction steel plate of not less

	than 5 mm in the floor and 3 mm to the sides, front and tailgate. The body is to have clear inside measurements of approximately 4 m × 2.5 m and to have a capacity of 12.5 m^3. The body is to have drop sides, middle and end removable support pillars, and a removable tailboard. The body should be treated with at least two coats of primer and finished in light orange high-gloss paint.
Fuel tank	The fuel tank should have a minimum capacity of 300 l and should be fitted with a lockable fuel cap.
Tachograph	The vehicle should be fitted with a tachograph. The vehicle must also be fitted with a speed limiter if the speed capability is greater than 52 mph.
Electrical system	The vehicle should be fitted with a 24 V electrical system for starting and lighting. An appropriately rated 24 V battery and alternator should be fitted. Efficient horn, head, side, tail and stop lamps are to be provided.

Carry out research using your school or college library and/or the Internet in order to answer the following questions:

1. Explain the meaning of the term *BHP*.
2. Explain why the vehicle needs to have a large number of gears.
3. Explain why power assisted steering is required.
4. What is a *leaf spring*?
5. What is a *shock absorber*?
6. What does *wheelbase* mean? Illustrate your answer with a sketch.
7. Describe THREE different safety features that must be incorporated.
8. What is a *tachograph* and why is it needed?
9. What is an *alternator* and why is it needed?
10. Draw a dimensioned sketch showing the body of the vehicle. Also determine the minimum height of the sides of the body in order to meet the client's requirements.

Engineering drawings are sometimes referred to as *formal* or *informal*. Informal drawings (see Figure 1.10(a)) are usually sketches or hand-drawn diagrams that provide a quick impression of what something will look like or how something will work. Formal drawings (see Figure 1.10(b)) generally take much longer to produce and usually contain a lot more detail. They are also much more precise and often include a scale and dimensions.

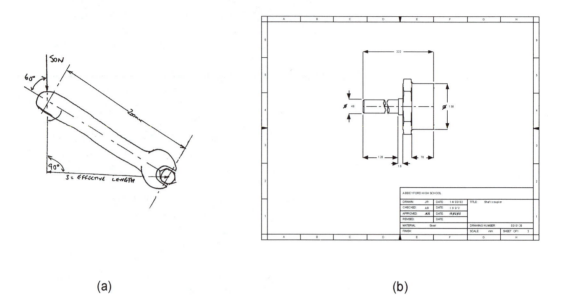

(a)   (b)

Figure 1.10 *Examples of informal (a) and formal (b) drawings*

1.3.1 Planning your drawing

Before you begin your drawing and put pencil to paper you should plan what you are going to do. This saves having to alter the drawing or even having to start again later on. You will have to decide whether the drawing is to be pictorial, orthographic or schematic. If orthographic you will have to decide on the projection you are going to use. You will also have to decide whether you need a formal drawing or whether a freehand sketch is all that is required. If a formal drawing is needed then you will have to decide whether to use manual drawing techniques or computer-aided design (CAD).

1.3.2 Paper size and orientation

When you start to plan your drawing you have to decide on the paper size that you are going to use. Engineering drawings are usually produced on 'A' size paper. Paper size A0 is approximately $1\,\text{m}^2$ in area and is the basis of the system. Size A1 is half the area of size A0, size A2 is half the area of size A1 and so on down to size A4. Smaller sizes are available but they are not used for drawing. All the 'A' size sheets have their sides in the ratio of 1:1.414. This

Test your knowledge 1.13

How many A5 sheets of paper can be cut from a single sheet of A3 drawing paper?

gives the following paper sizes:

A0 841 mm × 1189 mm
A1 594 mm × 841 mm
A2 420 mm × 549 mm
A3 297 mm × 420 mm
A4 210 mm × 297 mm

These relationships are shown diagrammatically in Figure 1.11.

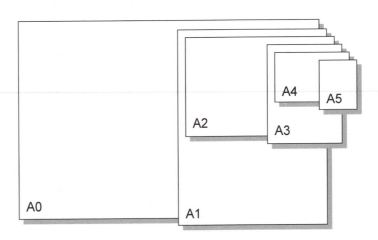
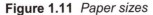

Figure 1.11 *Paper sizes*

Drawings can be arranged as either portrait or landscape view (see Figure 1.12). Landscape views are usually used for all formal engineering drawings. The paper size you choose will depend upon the size of the drawing and the number of individual drawings (or *views*) that might be required. Be generous, nothing looks worse than a cramped up drawing. It is also a false economy since overcrowding often leads to errors when a drawing is being read!

(a) Landscape   (b) Portrait

Figure 1.12 *Paper orientation*

1.3.3 Sketching

Sketching is one of the most useful tools available to the engineer to express his or her ideas and preliminary designs. Sketches are drawn freehand and they are used to gain a quick impression of what something will look like. A sketch can be either a two-dimensional (2D) representation (see Figure 1.13) or a three-dimensional (3D) representation (see Figure 1.14). A sketch can also take the form of a block diagram or a schematic diagram (see Figure 1.15). Where appropriate, labels, approximate dimensions and brief notes can be added to any of these types of drawing.

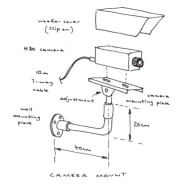

Figure 1.13 *A 2D sketch*

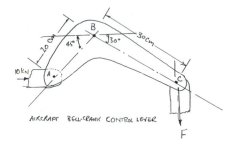
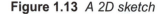
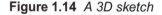

Figure 1.14 *A 3D sketch*

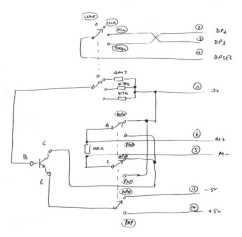
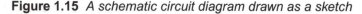

Figure 1.15 *A schematic circuit diagram drawn as a sketch*

Key point

Engineers use hand-drawn sketches to convey their ideas quickly and without having to use a lot of words. Sketches can be 2D or 3D drawings and may, or may not, include dimensions.

How to draw a sketch

✓ You will need an A3 or A4 sketch pad or a sheet of drawing or sketching paper, an HB pencil or a drawing pen and an eraser.

✓ Start by deciding how much space you will need for your sketch and how it will be positioned on the paper.

✓ Try to position the various parts of your sketch (and get the proportions right) by dividing up the sheet of paper in your mind using an invisible grid (if you find this hard you may find it easier to use paper that has been ruled with feint squares).

✓ Lightly pencil in each of the main features of your drawing ensuring that they are correctly positioned in relation to one another (you will find the grid useful for this).

✓ Continue until you have shown all of the features.

✓ Any lines that you are unhappy about can now be erased and redrawn.

✓ Label your sketch and/or add dimensions as required.

✓ Remove any lines or marks that you do not want to show in the finished sketch.

In order to make your sketch more useful you can

✓ Add shading where appropriate.

✓ Add notes to your drawing.

Test your knowledge 1.14

(a) Produce a 2D sketch of the open ended spanner shown in Figure 1.16.

(b) Produce a 3D sketch of the bicycle spanner shown in Figure 1.17.

Do not forget to add these sketches to your PDP.

Figure 1.16 *See Test your knowledge 1.14*

Figure 1.17 *See Test your knowledge 1.14*

Activity 1.9

Produce a 2D (plan-view) sketch of the printed circuit module shown in Figure 1.18. Add this sketch to your PDP.

Figure 1.18 *See Activity 1.9*

Activity 1.10

Produce a 3D sketch of a real car of your choice when viewed from the same direction as that shown in Figure 1.19. Indicate the make and model of the car and show this as a title below your sketch. Add this sketch to your PDP.

Figure 1.19 *See Activity 1.10*

Activity 1.11

A tubular steel mast has a diameter of 6 cm and a length of 4 m. The mast is to be secured to the ground at the base using a square steel plate with sides measuring 50 cm and by four equally spaced steel guy wires each having a length of 5 m anchored to the top of the mast. Produce a 3D sketch of this arrangement. Label your sketch and include dimensions. Add this sketch to your PDP.

1.3.4 Block diagrams

Block diagrams show the relationship between the various elements of a system. Figure 1.20 shows the block diagram of a simple radio receiver. This sort of diagram is often used in the initial stages of showing how a product or service will operate.

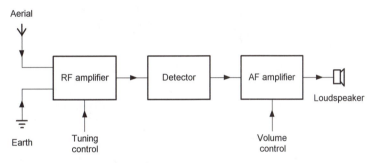

Figure 1.20 *A block diagram of a simple radio receiver*

Key point

Block diagrams are used to show how a number of things are connected together. Block diagrams use shapes (usually square or rectangular) connected together with arrowed lines.

How to draw a block diagram

✓ You will need an A3 or A4 sketch pad or a sheet of drawing or sketching paper, an HB pencil or a drawing pen and an eraser.

✓ Start by deciding how much space you will need for your block diagram and how it will be positioned on the paper.

✓ Try to position the various blocks by dividing up the sheet of paper in your mind using an invisible grid (if you find this hard you may find it easier to use paper that has been ruled with feint squares).

✓ Lightly pencil in each of the blocks in your block diagram and link them together with arrowed lines from the output of one block to the input of the next.

✓ Continue until you have shown all of the blocks.

✓ Check your diagram carefully. Any blocks or lines that you are unhappy about can be erased and redrawn.

✓ Label your block diagram and add a title.

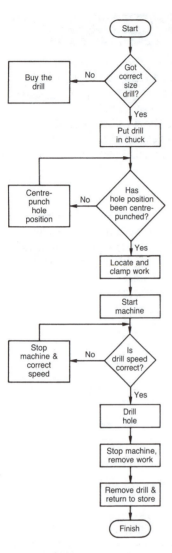

Figure 1.21 *Flow chart for drilling a hole*

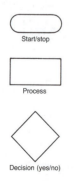

Figure 1.22 *Some standard flow chart symbols*

Activity 1.12

Construct a block diagram showing how the parts of an Internet-connected multimedia PC are connected together. Label your drawing clearly. Present your results in the form of a hand-drawn sketch and include it in your PDP.

1.3.5 Flow diagrams

Flow diagrams or flow charts are used to illustrate a sequence of events. They are used in a wide variety of applications including the planning of engineering processes and the design of computer software. Figure 1.21 shows a flow chart for the process of drilling a hole. The shapes of the symbols used in this flow chart have particular meanings as shown in Figure 1.22. For the complete set of symbols and their meanings you should refer to the appropriate British Standard, BS 4058.

How to draw a flow chart

✓ You will need an A3 or A4 sketch pad or a sheet of drawing or sketching paper, an HB pencil or a drawing pen and an eraser.

✓ Start by deciding how much space you will need for your flow chart and how it will be positioned on the paper.

✓ Try to position the various flow chart symbols by dividing up the sheet of paper in your mind using an invisible grid (if you find this hard you may find it easier to use paper that has been ruled with feint squares).

✓ Lightly pencil in each of the symbols in your flow chart and link them together with arrowed lines from one symbol to the next.

✓ Continue until you have shown all of the symbols.

✓ Check your flow chart carefully. Any blocks or lines that you are unhappy about can be erased and redrawn.

✓ Label your flow chart and add a title.

Activity 1.13

Your bicycle tyre is flat and may have a puncture or may simply need re-inflating. Draw a flow chart for checking the tyre and, if necessary, repairing or replacing it. Figure 1.23 will provide you with a starting point. Present your results in the form of a hand-drawn sketch and include it in your PDP.

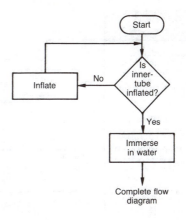

Figure 1.23 *See Activity 1.13*

1.3.6 Schematic diagrams

Schematic diagrams use standard symbols to show how things are connected together. There are several types of schematic diagram including those used for electrical and electronic circuits, pneumatic (compressed air), and hydraulic (compressed fluid) circuits.

Circuit diagrams are used to show the functional relationships between the components in an electric or electronic circuit. Components (such as resistors, capacitors, and inductors) are represented by symbols and the electrical connections between the components drawn using straight lines. It is important to note that the position of a component in a circuit diagram does not represent its actual physical position in the final assembly. Circuit diagrams are sometimes also referred to as *circuit schematics*.

Figure 1.24(a) shows the *circuit schematic* for a circuit using standard electronic component symbols. Figure 1.24(b) shows the corresponding physical *layout diagram* with the components positioned on the upper (component side) of a printed circuit board (PCB). Finally, Figure 1.24(c) shows the copper *track layout* for the PCB. This layout is developed photographically as an etch-resistant pattern on the copper surface of a copper-clad board.

A *wiring diagram* is another form of schematic diagram. It shows the *physical* connections between electrical and electronic components rather than the *electrical* connections between them. Schematic diagrams are also used to represent pneumatic (compressed air) circuits and hydraulic circuits. Pneumatic circuits and hydraulic circuits share the same symbols. You can tell which circuit is which because pneumatic circuits should have open arrowheads, whilst hydraulic circuits should have solid arrowheads. Also, pneumatic circuits exhaust to the atmosphere, whilst hydraulic circuits have to have a return path to an oil reservoir. Figure 1.25 shows a simple hydraulic circuit where the components are represented by standard symbols just as the electronic components were drawn in symbolic form in the circuit schematic shown in Figure 1.24(a).

Just as electrical and electronic circuit diagrams may have corresponding layout and wiring diagrams (see Figure 1.24), so do

Test your knowledge 1.15

(a) Refer to the circuit schematic shown in Figure 1.24. How many 741 integrated circuits are used?

(b) Refer to the component layout diagram shown in Figure 1.24(b). How many pins does a 741 integrated circuit have?

(c) Refer to the PCB track layout shown in Figure 1.24(c). How many pins does a 741 integrated circuit have?

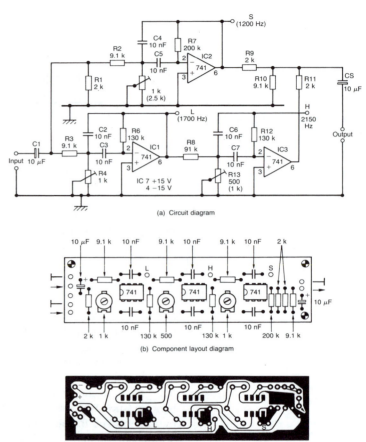

Figure 1.24 *Electronic circuit diagram together with corresponding component layout and PCB copper track layout diagrams*

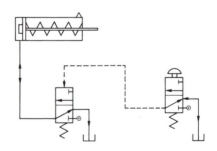

Figure 1.25 *A simple hydraulic circuit*

hydraulic, pneumatic and plumbing circuits. Only this time the wiring diagram becomes a *piping diagram*. A plumbing example is shown in Figure 1.26. As you may not be familiar with the symbols, I have named them for you. Normally this is not necessary and the symbols are recognised by their shapes.

Key point

Schematic diagrams are used to
show how components are
connected together in electrical,
pneumatic and hydraulic circuits.
Schematic diagrams use
standard symbols for
components and the links
between them are shown with
lines.

Key point

Just as electrical and electronic
circuit diagrams have
corresponding layout and wiring
diagrams, so do hydraulic,
pneumatic and plumbing circuits.
Physical layout diagrams show
how components and parts look
when they are assembled whilst
the wiring and piping diagrams
show how they are connected
together.

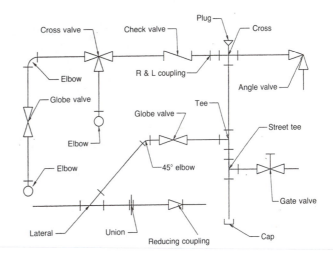

(a) Circuit diagram (schematic)

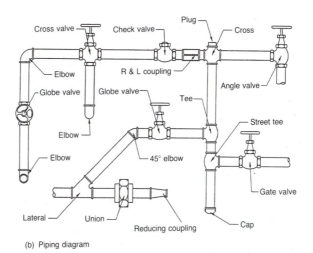

(b) Piping diagram

Figure 1.26 *A typical plumbing circuit with the corresponding
piping diagram*

How to draw a schematic diagram

✓ You will need an A3 or A4 sketch pad or a sheet of drawing
   or sketching paper, an HB pencil or a drawing pen and an
   eraser. You may also find it useful to have a set of standard
   symbols to refer to.

✓ Start by deciding how much space you will need for your
   schematic diagram and how it will be positioned on the paper.

✓ Try to position the schematic symbols by dividing up the
   sheet of paper in your mind using an invisible grid (if you
   find this hard you may find it easier to use paper that has
   been ruled with feint squares).

✓ Lightly pencil in each of the symbols in your schematic diagram flow chart and link them together with lines from one symbol to the next.

✓ Continue until you have shown all of the symbols.

✓ Check your diagram carefully. Any symbols or lines that you are unhappy about can be erased and redrawn.

✓ Label your schematic diagram and add a title.

1.3.7 Laying out formal drawings

Figure 1.27 shows the layout of a typical formal drawing. A formal engineering drawing should always have a border and a title block. So that individual features of a drawing can be easily located, the border often has letters along one axis and numbers along the other. It is thus possible to identify a particular *drawing zone*, for example C4 has been shaded in Figure 1.27. To save time, formal drawing sheets are usually printed to a standard layout for use in a particular company ready for the draftsperson to add the drawing and complete the boxes and tables. A selection of formal drawing sheets are also provided as *templates* in most CAD programs.

The title block can be expanded horizontally to accommodate any written information that is required. The title block should contain:

- The drawing number (which is often repeated in the top left-hand corner of the drawing)
- The drawing name or title
- The scale used for the drawing
- The projection used (first or third angle)
- The name and signature of the person who made the drawing (i.e. the *originator*)
- The name and signature of the person who checks and/or approves the drawing, together with the date
- The drawing issue number and its release date
- Any other information as required.

In addition, a list of component parts may be provided (together with numbered references shown on the drawing), the materials that are to be used, the finish that is to be applied, the units used for measurement and tolerances, reference to appropriate standards (e.g. BS 308), and guidance notes (such as 'do not scale').

The scale should be stated on the drawing as a ratio. The recommended scales are as follows:

- Full size, that is 1:1
- Reduced scales (smaller than full size), for example: 1:2, 1:5, 1:10, 1:20, 1:50, 1:100, 1:200, 1:500, 1:1000. Do not use the words full size, half size, quarter size and so on!
- Enlarged scales (larger than full size), for example: 2:1, 5:1, 10:1, 20:1, 50:1, 100:1.

Key point

Formal engineering drawings should always have a border and a title block. The border may have letters and numbers so that drawing zones can be easily located. The title block contains important information such as title, scale, the company name and the name of the originator.

Test your knowledge 1.17

List five items that usually appear in the title block of a formal engineering drawing.

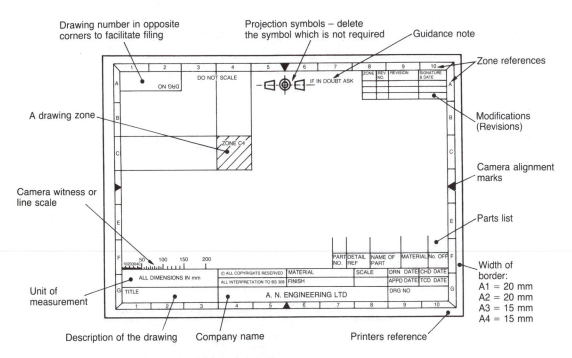

Figure 1.27 *Layout of a typical formal drawing*

1.3.8 General arrangement drawings

Figure 1.29 shows a typical *general arrangement* (GA) drawing. This shows as many of the features listed above as are appropriate for this drawing. It shows all the components correctly assembled together. Dimensions are not usually given on GA drawings although, sometimes, overall dimensions will be given for reference when the GA drawing is of a large assembly drawn to a reduced scale.

The GA drawing shows all the parts used in an *assembly*. These are often listed in a table together with the quantities required. Manufacturers' catalogue references are also given for any components that are not actually being manufactured. The parts are usually 'bought-in' as 'off-the-shelf' parts from other suppliers. The detail drawing numbers are also included for components that have to be manufactured as special items.

Key point

GA drawings show an engineered product that is assembled from a number of individual parts. Some of these parts may have to be produced whilst others may be available 'off-the-shelf' from other suppliers or manufacturers.

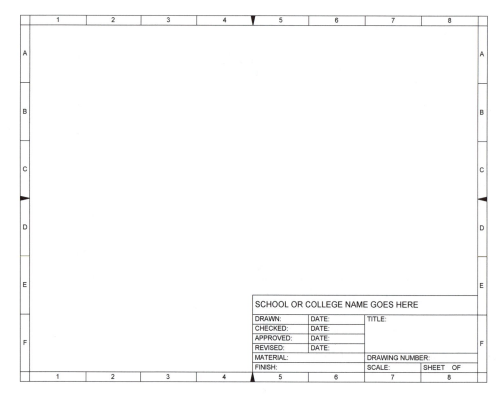

Figure 1.28 *See Activity 1.14*

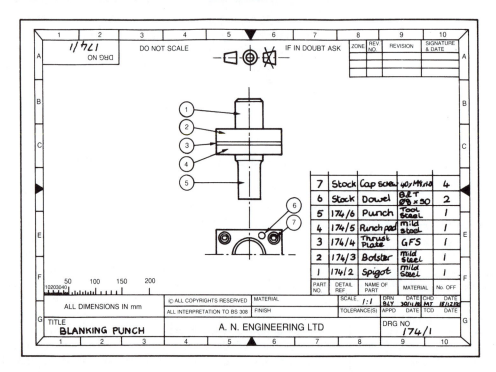

Figure 1.29 *A GA drawing*

1.3.9 Detail drawings

As the name implies, detail drawings provide all the details required to make the component shown on the drawing. If you look back at Figure 1.29 you will see from the table that the detail drawing for the punch has the reference number 174/6. Figure 1.30 shows this detail drawing. In this instance, the drawing provides the following information:

- The shape of the punch
- The dimensions of the punch and the manufacturing tolerances
- The material from which the punch is to be made and its subsequent heat treatment
- The unit of measurement (millimetre)
- The projection (first angle)
- The finish
- The guidance note 'Do not scale drawing'
- The name of the Company
- The name of the draftsperson
- The name of the person checking the drawing.

It should go without saying that the amount of information given will depend upon the nature of the job. Drawings for a critical aircraft component, for example, will be much more fully detailed than those for a wheelbarrow component!

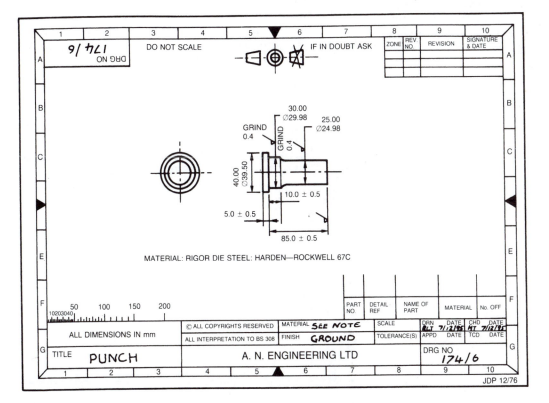

Figure 1.30 *A typical detail drawing*

Test your knowledge 1.19

Refer to the detail drawing shown in Figure 1.30.

(a) What units are used for the dimensions?
(b) What is the overall length of the component?
(c) What is the diameter of the component?
(d) What finish is applied to the component?
(e) What material is used for the component?

Key point

A detail drawing shows all information necessary to determine the final form of a part whether it is to be manufactured or 'bought-in'. The detail drawing must show a complete and exact description of the part including dimensions, tolerances, materials and finish.

Activity 1.15

Use one of the blank formal drawing sheets that you created in Activity 1.14 to create your own copy of Figure 1.30. Draw this actual size (i.e. a 1:1 scale). Do not forget to complete the title block and include the drawing in your PDP.

1.3.10 Using CAD

CAD has now largely replaced manual methods used for producing formal engineering drawing. CAD software is used in conjunction with a computer and the drawing produced on the computer screen is saved in a computer file on disc. Networked CAD/CAM (Computer-aided manufacture) and computer-aided engineering (CAE) systems have made it possible to share data and drawings over a network and also make them available to computer numerically controlled (CNC) machine tools that carry out automated manufacturing operations. Your school or college will be able to provide you with access to CAD equipment as part of your GCSE course. Figures 1.31–1.33 show screen shots of three popular CAD packages.

Figure 1.31 shows a simple 2D drawing of a gasket using DesignCAD. The drawing is made to scale and dimensions are still to be added. Figure 1.32 shows a more complicated GA drawing produced using AutoSketch. Some dimensions have been included. Finally, Figure 1.33 shows a 3D assembly diagram produced using AutoCAD. Drawings like this show how a complex product is assembled from its component parts. Product design involves a great deal of 3D work like this in the initial stages of designing a product before any of the parts or components are manufactured.

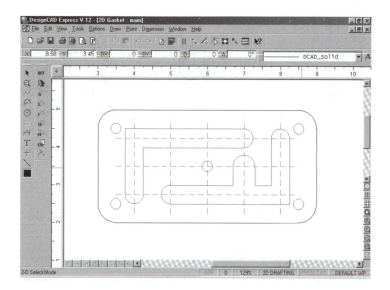

Figure 1.31 *A 2D drawing of a gasket using DesignCAD*

Key point

CAD refers to the use of a computer to design a part and to produce engineering drawings. 2D CAD is used to show schematics, GA and detail drawings (Figure 1.34). 3D CAD is often used for assembly diagrams and in product design (Figure 1.35).

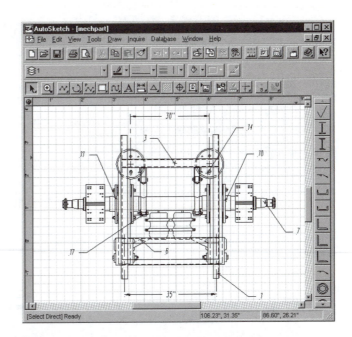

Figure 1.32 *A GA drawing using AutoSketch*

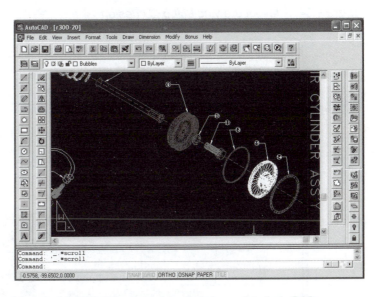

Figure 1.33 *A 3D assembly diagram using AutoCAD*

Activity 1.16

Repeat Activity 1.14 but use CAD to produce the formal engineering drawing template. Once again, include your school name on the template and your initials. Save your completed template on a floppy disc (or in your own directory if you are using a networked system) and make a printed hard copy on an A4 sheet to include in your PDP.

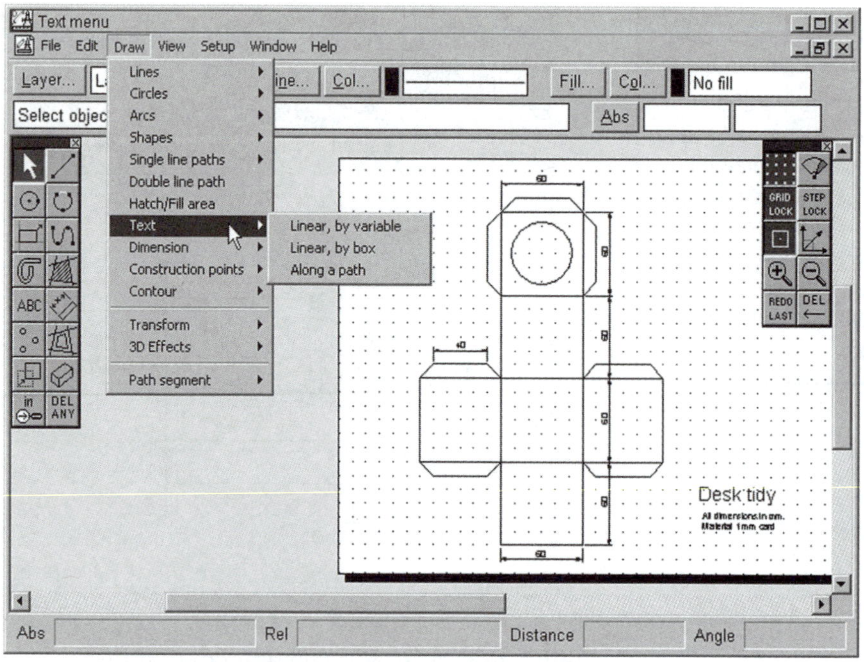

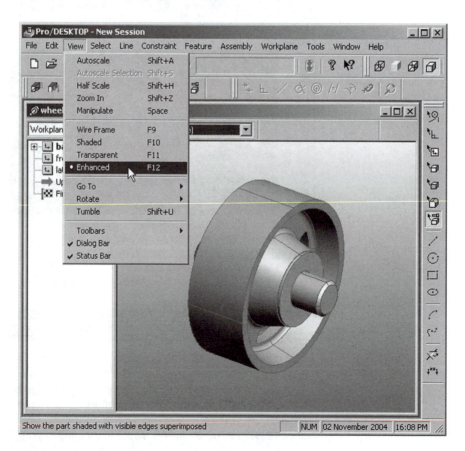

Figure 1.34 *A layout drawing using 2D design*

Figure 1.35 *A rendered CAD drawing using ProDesktop*

1.3.11 Drawing standards

To avoid confusion, engineering drawings make use of nationally and internationally recognised symbols, conventions and abbreviations which are explained in the appropriate British Standards. Standard conventions are used in order to avoid having to draw out, in detail, common features in frequent use. Figure 1.36 shows a typical dimensioned engineering drawing. Some conventions can help save you a great deal of time and effort. For example, Figure 1.37(a) shows a pictorial representation of a screw thread, whilst Figure 1.37(b) shows the standard convention for a screw thread that is much quicker and easier to draw!

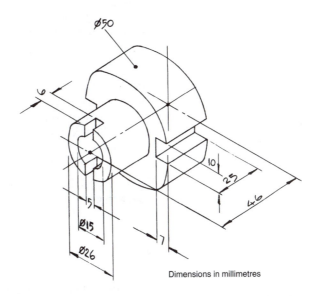

Dimensions in millimetres

Figure 1.36 *A typical dimensioned engineering drawing*

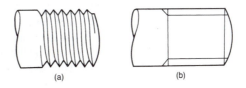

(a)   (b)

Figure 1.37 *A screw thread (a) and corresponding drawing convention (b)*

Test your knowledge 1.20

What do each of the following British Standards relate to?

(a) PP 8888
(b) BS 3939.

Test your knowledge 1.21

Refer to the detail drawing shown in Figure 1.30. Which British Standards is referred to?

If you need further information on drawing standards you should, in the first instance, take a look at the British Standards Institution's publication PP 8888, *Engineering Drawing Practice for Schools and Colleges*. This document provides some useful information that will help you improve your drawings and a copy should be available in your school or college.

Other British Standards of importance to engineering draftspersons and designers are:

- BS 308 Engineering Drawing Practice
- BS 4500 ISO Limits and Fits (these are used by mechanical and production engineers)
- BS 3939 Graphical symbols for electrical power, telecommunications and electronics diagrams
- BS 2197 Specifications for graphical symbols used in diagrams for fluid power systems and components
- PP 7307 Graphical symbols for use in Schools and Colleges.

Activity 1.18

In this activity you are required to use appropriate CAD techniques to illustrate a design for a simple Desk Tidy. The Desk Tidy is to be manufactured from pre-printed 1 mm card (see Figure 1.38) which is cut and folded to produce the finished container shown in Figure 1.39.

Use appropriate techniques and CAD packages (such as 2D Design or ProDesktop) to produce the following:

(a) a hand-drawn sketch of the Desk Tidy (based on Figure 1.39)
(a) a fully dimensioned layout drawing for the Desk Tidy (like that shown in Figure 1.38)
(b) a solid 3D model of the Desk Tidy (like that shown in Figure 1.40).

Save and print your completed drawings and add them to your PDP.

1.3.12 Lines

The lines used in a drawing should be uniformly black, dense and bold. On any one drawing they should all be in pencil or in black ink. Pencil is quicker to use but ink prints out more clearly. Lines should be thick or thin as recommended below. Thick lines will normally be twice as thick as thin lines. Figure 1.41 shows the types of lines recommended in BS 308 for use in engineering drawing and what they should be used for.

Sometimes the lines overlap in different views. When this happens the following order of priority should be observed:

- Visible outlines and edges (type A) take priority over all other lines
- Next in importance are hidden outlines and edges (type E)

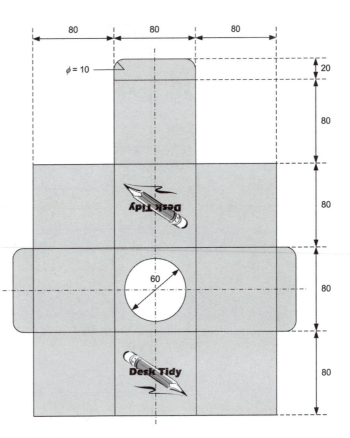

80   80   80

$\phi = 10$

20

80

80

80

60

80

All dimensions in mm

Figure 1.38 *Layout drawing for the Desk Tidy (prior to cutting and folding)*

Figure 1.39 *Final appearance of the completed Desk Tidy*

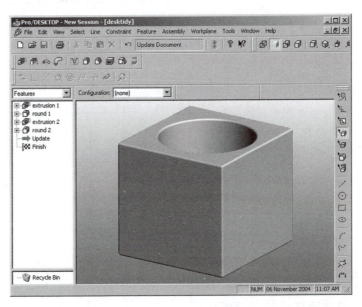

Figure 1.40 *Using ProDesktop to produce a 3D model of the Desk Tidy (which can be revolved and rotated on-screen)*

Line type	Example	Description	Application
A	 ————	Continuous thick	Visible outlines and edges
B	 ————	Continuous thin	Dimension, projection and leader lines, hatching, outlines of revolved sections, short centre lines, imaginary intersections
C	 ∿∿∿	Continuous thin irregular	Limits of partial or interrupted views and sections
D	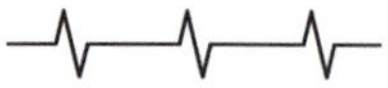 ⟋∿⟋∿⟋	Continuous thin with zigzags	Limits of partial or interrupted views and sections
E	 – – – – –	Dashed thin	Hidden outlines and edges
F	 – · – · – · –	Chain thin	Centre lines, lines of symmetry, trajectories and loci, pitch lines and pitch circles
G	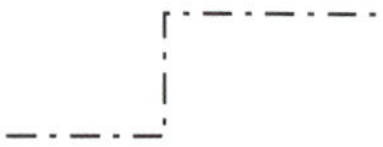 ⌐ · – · – · ⌐	Chain thin, thick at ends and changes of direction	Cutting planes
H	 – · · – · · –	Chain thin double dashed	Outlines and edges of adjacent parts, outlines and edges of alternative and extreme positions of movable parts, initial outlines prior to forming, bend lines on developed blanks or patterns

Figure 1.41 *Types of lines*

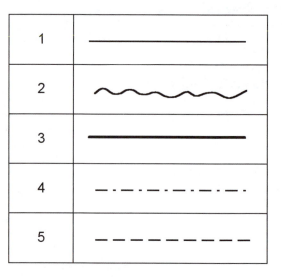

Figure 1.42 *See Test your knowledge 1.22*

Test your knowledge 1.22

Identify each of the line styles shown in Figure 1.42.

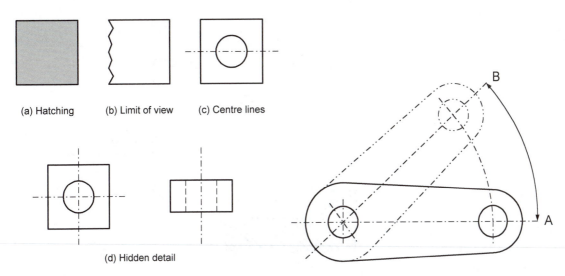

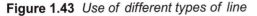

Figure 1.43 *Use of different types of line*

Figure 1.44 *Indicating extreme positions*

- Then cutting planes (type G)
- Next come centre lines (type F and B)
- Outlines and edges of adjacent parts, etc. (type H)
- Projection lines and shading lines (type B)

Figures 1.43 and 1.44 show some examples of how the rules concerning lines should be applied:

- In Figure 1.43(a), hatching (type B) is used inside a continuous area (type A).

- In Figure 1.43(b), the limit of view is indicated (type C).

- In Figure 1.43(c), the centre lines have been marked (type F).

- In Figure 1.43(d), the hidden detail is shown (type E) when the part is viewed from the side (note that the left-hand diagram is a plan view whilst the right-hand diagram is the corresponding side view).

- In Figure 1.44, the component shown is able to move from its resting position A to position B. Its extreme position is shown using line type H.

Leader lines are used when written information or dimensions need to be added to a drawing. Leader lines are thin lines (type B) and they end in an arrowhead or in a dot as shown in Figure 1.45(a).

Arrowheads touch and stop on a line, whilst dots should always be used *within* an outline. The rules for arrowheads are as follows:

- When an arrowed leader line is applied to an arc it should be in line with the centre of the arc as shown in Figure 1.45(b).

- When an arrowed leader line is applied to a flat surface, it should be nearly normal to the lines representing that surface as shown in Figure 1.45(c).

- Long and intersecting leader lines should not be used, even if this means repeating dimensions and/or notes as shown in Figure 1.45(d).

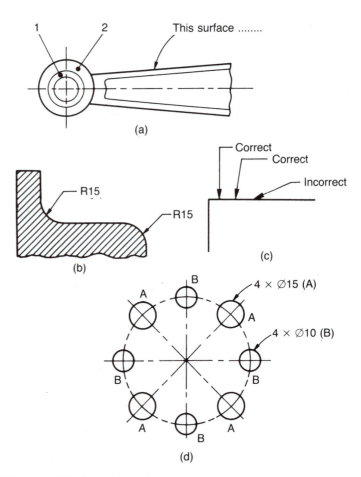

Figure 1.45 *Examples of the use of leader lines*

- Leader lines must not pass through the points where other lines intersect.

- Arrowheads should be triangular with their length some three times larger than the maximum width. They should be formed from straight lines and the arrowheads should be filled in. The arrowhead should be symmetrical about the leader line, dimension line or stem. It is recommended that arrowheads on dimension and leader lines should be some 3–5 mm long.

- Arrowheads showing direction of movement or direction of viewing should be some 7–10 mm long. The stem should be the same length as the arrowhead or slightly greater. It must never be shorter.

1.3.13 Letters and numbers

The following conventions apply to the use of letters and numbers in formal engineering drawings.

Test your knowledge 1.23

Figure 1.46 shows some applications of leader lines with arrowheads and leader lines with dots. List the numbers and state whether the application is correct or incorrect. If incorrect explain (with sketches if required) how the application should be corrected.

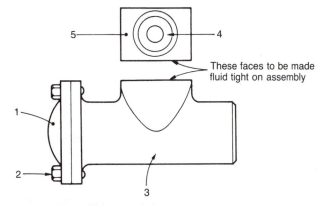

Figure 1.46 *See Test your knowledge 1.23*

Style
The style should be clear and free from embellishments. In general, capital letters should be used. Suitable styles are shown in Figure 1.47.

Size
The characters used for dimensions and notes on drawings should not be less than 3 mm tall. Title and drawing numbers should be at least twice as big.

Direction of lettering
Notes and captions should be positioned so that they can be read in the same direction as the information in the title block. Dimensions have special rules and will be dealt with later.

Location of notes
General notes should all be grouped together and not scattered about the drawing. Notes relating to a specific feature should be placed adjacent to that feature.

Emphasis
Characters, words and/or notes should not be emphasised by underlining. Where emphasis is required the characters should be enlarged.
Finally, please remember that, in order to save space, symbols and abbreviations are frequently used in formal engineering drawings. You should refer to BS 308 or PP 8888 if you need further information.

Test your knowledge 1.24

Which of the following statements are TRUE when adding letters and numbers to an engineering drawing?

(a) Only lower-case letters should be used.
(b) Words should be emphasised by using underlining.
(c) The same size should be used for all of the letters and numbers.
(d) The style used for letters and numbers should be as simple as possible.
(e) Notes relating to a particular feature should be placed in the title block.

ABCDEFGHIJKLMNOPQRSTUVWXYZ
1234567890

ABCDEFGHIJKLMNOPQRSTUVWXYZ
1234567890

ABCDEFGHIJKLMNOPQRSTUVWXYZ
1234567890

ABCDEFGHIJKLMNOPQRSTUVWXYZ
1234567890

Figure 1.47 *Suitable styles for letters and numbers in engineering drawings*

1.3.14 Drawing conventions

We have already seen how standard drawing conventions can help save time and effort when drawing a screw thread (take a look back at Figure 1.37). Conventions are a form of 'shorthand' that is used to speed up the drawing of common features that are in regular use. The full range of conventions and examples of their use can be found in appropriate standards so we will not waste space by listing them here. However by completing the next exercise you will use some of the more common conventions and this will help you to become familiar with them.

Activity 1.19

Copy Figure 1.48 to an A4 sheet of drawing paper. With reference to appropriate standards (your teacher or tutor will provide you with these) complete the blank spaces. Note that you must take care to use the same types of line as shown in the standard or the conventions become meaningless (this applies particularly to line thickness). Include your finished drawing in your PDP.

1.3.15 Dimensions

When a component is being dimensioned, the dimension lines and the projection lines should be thin full lines (type B). Where possible dimensions should be placed outside the outline of the object as shown in Figure 1.49(a). The rules are:

Key point

A *dimension* is a numeric value expressed in appropriate units of measure and indicated on a drawing and in other documents along with lines, symbols, and notes to define the size or geometric characteristic, or both, of a part or part feature. A *reference dimension* (usually specified without a tolerance) is used for information only. A reference dimension repeats a dimension or size already given or derived from other values shown on the drawing or related drawing. Reference dimensions are enclosed in brackets, for example (23.50).

- Outline of object to be dimensioned in thick lines (type A)
- Dimension and projection lines should be half the thickness of the outline (type B)
- There should be a small gap between the projection line and the outline
- The projection line should extend to just beyond the dimension line
- Dimension lines end in an arrowhead that should touch the projection line to which it refers
- All dimensions should be placed in such a way that they can be read from the bottom right-hand corner of the drawing.

The purpose of these rules is to allow the outline of the object to stand out prominently from all the other lines and to prevent confusion.

There are three ways in which a component can be dimensioned. These are:

- Chain dimensioning as shown in Figure 1.49(b)
- Absolute dimensioning (dimensioning from a datum) using parallel dimension lines as shown in Figure 1.49(c)
- Absolute dimensioning (dimensioning from a datum) using superimposed running dimensions as shown in Figure 1.49(d). Note the common origin (termination) symbol.

A *tolerance* is the total amount by which a specific dimension is permitted to vary. The tolerance is the difference between the maximum and the minimum limits.

The *origin* or *datum point* is the name given to the point from where the location or geometric characteristics of a part are established. The correct identification of datums on a component and the related dimensioning can be vitally important in the manufacturing process (e.g. CNC machining).

A *feature* is a general term applied to a physical portion of a part, for example a surface, hole or slot. A *datum feature* is a geometric feature of a part that is used to establish a datum. For example a point, line, surface, hole, etc.

Explain each of the following terms:

(a) Reference dimension
(b) Tolerance
(c) Datum point
(d) Datum feature.

TITLE	SUBJECT	CONVENTION
External screw threads (details)		
Screw threads (assembly		
Compression springs		
Diamond knurling		
Square on shaft		
Holes on linear pitch		

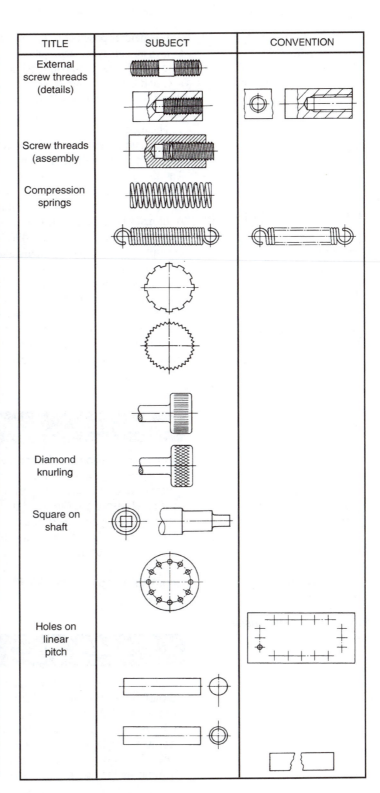

Figure 1.48 *See Activity 1.19*

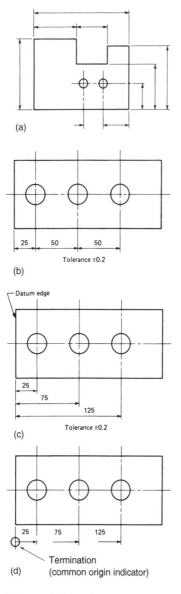

(a)

(b)

Tolerance ±0.2

25   50   50

Datum edge

(c)

Tolerance ±0.2

25
75
125

(d)

25   75   125

Termination
(common origin indicator)

Figure 1.49 *Dimensioning*

It is neither possible to manufacture an object to an exact size nor to measure an exact size. Therefore, important dimensions have to be *toleranced*. That is, the dimension is given two sizes; an upper limit of size and a lower limit of size. Providing the component is made so that it lies between these limits it will function correctly. Information on Limits and Fits can be found in BS 4500.

The method of dimensioning can also affect the accuracy of a component and produce some unexpected effects. Figure 1.49(b) shows the effect of chain dimensioning on a series of holes or other features. The designer specifies a common tolerance of ± 0.2 mm. However, since this tolerance is applied to each and every dimension, the cumulative tolerance becomes ± 0.6 mm by the time you reach the final, right-hand hole. Not what was intended! Therefore, absolute dimensioning as shown in Figure 1.49(c) and (d) is to be preferred in this example. With absolute dimensioning, the position of each hole lies within a tolerance of ± 0.2 mm and there is no cumulative error.

Reference dimensions are usually specified without a tolerance. They are used for information only and not for production or inspection purposes. A reference dimension repeats a dimension or size already given or derived from other values shown on the drawing or related drawing. Reference dimensions are enclosed in brackets, for example (23.50). Further examples of dimensioning techniques are shown in Figure 1.50.

Activity 1.20

Figure 1.51 shows a drawing with some leader lines and dimensions added. Each of the circular holes is to have a diameter of 20 mm and is to be centred 20 mm from the edge. Redraw the diagram by applying CAD or manual techniques (your teacher or tutor will tell you which to use) using one of the drawing templates that you created in Activity 1.13 or 1.15. Add the remaining dimensions using correct drawing conventions. When your drawing is complete add it to your PDP.

Activity 1.21

Figure 1.52 shows an aluminium plate which has dimensions 140 cm × 40 cm. The three holes each have a diameter of 20 cm and their centres are spaced by 50 cm. Choose a suitable scale and redraw the drawing by applying CAD or manual techniques (your teacher or tutor will tell you which to use) using one of the drawing templates that you created in Activity 1.13 or 1.15. Include a full set of dimensions using appropriate drawing conventions. When your drawing is complete add it to your PDP.

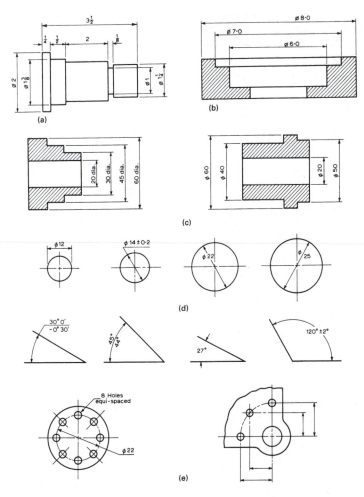

Figure 1.50 *Further examples of dimensioning*

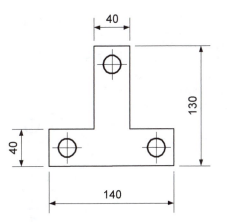

Figure 1.51 *See Activity 1.20*

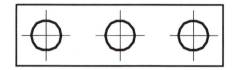

Figure 1.52 *See Activity 1.21*

1.3.16 Sections and sectional views

Sections are used to show the hidden detail inside hollow objects more clearly than can be achieved using dashed thin (type E) lines. Figure 1.53 shows an example of a simple sectioned drawing. The *cutting plane* is the line A–A. In your imagination you remove everything to the left of the cutting plane, so that you only see what remains to the right of the cutting plane looking in the direction of the arrowheads. Another example is shown in Figure 1.54.

Figure 1.55 shows how to section an assembly. Note how solid shafts and the key are not sectioned. Also note that thin webs that lie on the section plane are not sectioned. When interpreting sectioned drawings, some care is required!

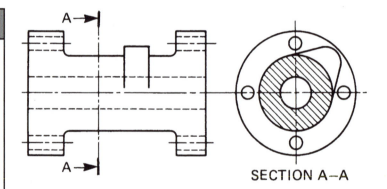

Figure 1.53 *A simple sectioned drawing*

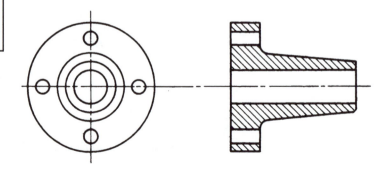

Figure 1.54 *Another example of a sectioned drawing*

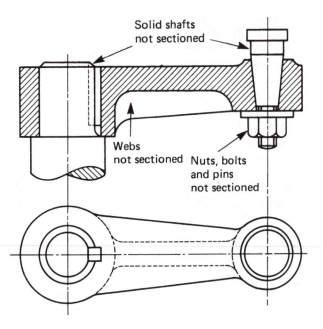

Figure 1.55 *A sectioned assembly*

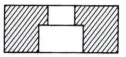

(a)

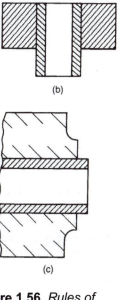

(b)

(c)

Figure 1.56 *Rules of hatching*

You will have noticed that the shading of sections and sectional views consists of sloping, thin (type B) lines. This is called *hatching*. The lines are equally spaced, slope at 45° and are not usually less than 4 mm apart. However, when hatching very small areas the hatching can be reduced, but never less than 1 mm. The drawings in this book may look as though they do not obey these rules. Remember that they have been reduced from much bigger drawings to fit onto the pages.

Figure 1.56 shows the basic rules of hatching. The hatching of separated areas is shown in Figure 1.56(a). Separate sectioned areas of the same component should be hatched in the same direction and with the same spacing.

Figure 1.56(b) shows how to hatch assembled parts. Where the different parts meet on assembly drawings, the direction of hatching should be reversed. The hatching lines should also be staggered. The spacing may also be changed.

Figure 1.56(c) shows how to hatch large areas. This saves time and avoids clutter. The hatching is limited to that part of the area that touches adjacent hatched parts or just to the outline of a large part.

1.4 Engineering drawing techniques

Engineering drawings can be produced using a variety of different techniques. The choice of technique is dependent upon a number of factors as follows.

Speed
How much time can be allowed for producing the drawing. How soon the drawing can be commenced.

Media

The choice will depend upon the equipment available (e.g. CAD or conventional drawing board and instruments) and the skill of the person producing the drawing.

Complexity

The amount of detail required and the anticipated amount and frequency of development modifications.

Cost

Engineering drawings are not works of art and have no intrinsic value. They are only a means to an end and should be produced as cheaply as possible. Both initial and ongoing costs must be considered.

Presentation

This will depend upon who will see/use the drawings. Non-technical people can visualise pictorial representations better than orthographic drawings.

Nowadays engineering drawings are increasingly produced using CAD techniques. Developments in software and PCs have reduced the cost of CAD and made it more powerful. At the same time, it has become more 'user friendly'.

CAD does not require the high physical skill required for manual drawing that often takes years of practice to achieve. CAD also has a number of other advantages over manual drawing. Let us consider some of these advantages.

Accuracy

Points and dimensions can be fixed very accurately and do not depend on the draftsperson.

Radii

Radii can be made to blend with straight lines automatically.

Repetitive features

For example, holes round a pitch circle do not have to be individually drawn but can be easily produced automatically by 'mirror imaging'. Again, some repeated, complex feature need only be drawn once and saved as a matrix. It can then be called up from the computer memory at each point in the drawing where it appears at the touch of a key.

Editing

Every time you erase and alter a manually produced drawing the surface of the drawing is increasingly damaged. On a computer you can delete and redraw as often as you like with no ill effects.

Storage

No matter how large and complex the drawing, it can be copied, stored and transmitted digitally with no loss of detail.

Prints

Hard copy can be produced accurately and easily on laser printers, flat bed or drum plotters and to any scale. Colour prints can also be made.

1.4.1 Orthographic projection

When you produced the simple GA and detail drawings in 1.3 you made use of a basic drawing technique called *orthographic projection*. Orthographic projection is used to represent 3D solids on the 2D surface of a sheet of drawing paper so that all the dimensions are true length and all the surfaces are true to shape. To achieve this when surfaces are inclined to the vertical or the horizontal we have to make use of additional *auxiliary views*, but more about these later. Let us keep things simple for the moment.

Figure 1.57 shows a brass component used in the waveguide assembly of a high-power radar. Figure 1.58 shows how this component is drawn (together with dimensions) using conventional orthographic drawing techniques. On its own this drawing does not convey enough information for anyone to actually manufacture or specify the component from an off-the-shelf supplier. At this point you might like to think about what is missing and how you could show the additional information that is needed!

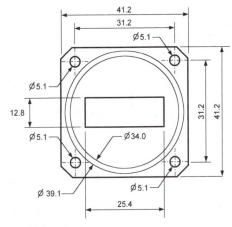

Figure 1.57 *A brass component used in the waveguide assembly of a high-power radar*

All dimensions in mm

Figure 1.58 *A conventional orthographic drawing of the component shown in Figure 1.55*

Key point

Orthographic projection is used to represent 3D solids on the 2D surface of a sheet of drawing paper so that all the dimensions are true length and all the surfaces are true to shape.

Test your knowledge 1.27

Refer to Figures 1.57 and 1.58.

(a) What are the external dimensions of the component?
(b) What are the internal dimensions of the component?
(c) What is the inner diameter of the circular groove in the face of the component?
(d) What is the outer diameter of the circular groove in the face of the component?
(e) What is the diameter of the component's four locating holes?

1.4.2   Oblique drawing

The most obvious way of improving Figure 1.58 is to attempt to draw the component in *perspective* (i.e. in a 3D view). The easiest way of doing this is to use a simple pictorial technique called *oblique drawing*. Figure 1.59 shows a simple oblique drawing using both *cavalier oblique* and *cabinet oblique* projection.

It is important to note that regardless of whether cavalier oblique or cabinet oblique projection is used, the front view (called the *elevation*) is drawn true shape and size. Therefore this view should be chosen so as to include any circles or arcs so that these can be drawn with compasses when using manual drawing techniques or using the circle drawing tool when using CAD.

The lines forming the side views appear to travel away from you, so these are called *receders*. They are drawn at 45° to the horizontal using a 45° set square if you are using manual drawing techniques. They may be drawn full length if you are using *cavalier oblique* drawing or they may be drawn half length if you are using *cabinet oblique* drawing. This latter method gives a more realistic representation, and is the one we will be using.

Key point

In oblique projection the *receders* are drawn at 45° to the horizontal. In cavalier *oblique projection* the receders are drawn full length whilst in *cabinet oblique* projection they are drawn half length.

Test your knowledge 1.28

Explain, with the aid of a sketch, the difference between cavalier oblique and cabinet oblique projection.

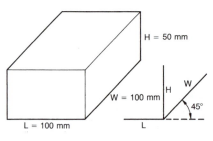

Cavalier oblique projection

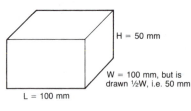

Cabinet oblique projection

Figure 1.59 *A simple oblique drawing*

How to construct an oblique drawing

✓ You will need an A3 or A4 sketch pad or a sheet of drawing or sketching paper, an HB pencil or a drawing pen and an eraser. Alternatively, you will need access to a CAD system.

✓ Start by deciding how much space you will need for your oblique drawing. What scale you will use, and how your drawing will be positioned on the drawing sheet.

✓ Construct the elevation. This is drawn using its true shape but scaled appropriately.

✓ Next add the receders at an angle of 45° to the horizontal. The receders need to be drawn at their full scaled size for cavalier oblique projection but they should be scaled by a further 50% of their scaled size for cabinet oblique projection.

✓ Complete the drawing showing all features and add labels and dimensions as required.

✓ Continue until you have shown all of the features.

✓ Any lines that you are unhappy about can be erased and redrawn.

✓ Complete the title block.

In order to make your oblique drawing more useful you can

✓ Add a note saying which projection was used for the drawing.

1.4.3 Isometric drawing

Isometric drawing provides us with another way of showing a 3D pictorial view of an object. Figure 1.60 shows an isometric view of the same box that we used for our examples of cavalier and cabinet oblique projection. You should take a little time to compare Figures 1.59 and 1.60 until you understand what the differences are!

When constructing the isometric view, the vertical lines should be drawn true length and the receders should be drawn to a special isometric scale. However this sort of accuracy is rarely required and, for all practical purposes, we can usually draw all the lines full size. As you can see, the receders are drawn at 30° to the horizontal for both the elevation and the end view.

Although an isometric drawing is more pleasing to the eye, it has the disadvantage that all circles and arcs have to be constructed and cannot simply be drawn with compasses. Special isometric drawing paper is available if your find that you have to produce isometric drawings on a regular basis.

> **Key point**
>
> In *isometric projection* the *receders* are normally drawn full size and at 30° to the horizontal.

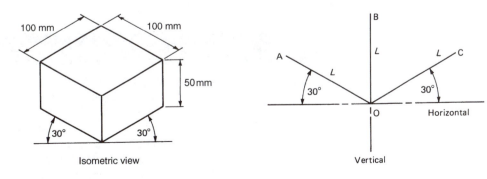

Figure 1.60 *An isometric view*

Test your knowledge 1.29

Explain, with the aid of a sketch, the difference between isometric and oblique drawing techniques.

How to construct an isometric drawing

✓   You will need an A3 or A4 sketch pad or a sheet of drawing or sketching paper, an HB pencil or a drawing pen and an eraser. Alternatively, you will need access to a CAD system.

✓   Start by deciding how much space you will need for your isometric drawing. What scale you will use, and how your drawing will be positioned on the drawing sheet.

✓   Construct the front elevation. The receders are drawn scaled from full size at an angle of 30° to the horizontal.

✓   Next construct the end elevation. Once again, the receders are drawn scaled from full size at an angle of 30° to the horizontal (*but on the other side* from that which you used for the front elevation).

✓   Complete the drawing showing all features and add labels and dimensions as required.

✓   Any lines that you are unhappy about can be erased and redrawn.

✓   Complete the title block.

In order to make your isometric drawing more useful you can

✓   Add a note saying which projection was used for the drawing.

Activity 1.22

Figure 1.61 shows a component used in a microwave radio system. Sketch this component in:

(a)   cabinet oblique view
(b)   isometric view.

There is no need to draw the component accurately but your sketches MUST show the correct use of the drawing projection in each case. Label your sketches and add them to your PDP.

Activity 1.23

Figure 1.62 shows a component drawn in isometric view. Redraw this component using cabinet oblique projection and include dimensions. You should use either manual or CAD drawing techniques as instructed by your teacher or tutor. In either case you should make use of one of your blank drawing sheet templates. Complete the title block and add the finished drawing to your PDP.

Figure 1.61 *See Activity 1.22*

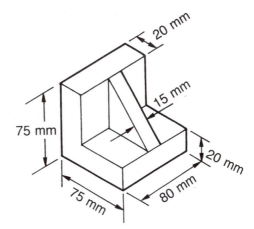

Figure 1.62 *See Activity 1.23*

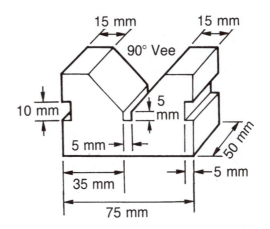

Figure 1.63 *See Activity 1.24*

> ### Activity 1.24
>
> Figure 1.63 shows a component drawn in cabinet oblique view. Redraw this component using isometric projection and include dimensions. You should use either manual or CAD drawing techniques as instructed by your teacher or tutor. In either case you should make use of one of your blank drawing sheet templates. Complete the title block and add the finished drawing to your PDP.

1.4.4   First-angle projection

Both oblique and isometric drawing projections provide a good way of showing the general features of a component. An alternative to using these techniques is providing a series of orthographic views. Often, just three views are required to show all the details required for manufacture or production of a part. Engineers use two orthographic drawing techniques called *first-angle projection* and *third-angle projection*. The former is called 'English projection' and the latter is called 'American projection'. We shall begin by looking at first-angle projection.

Figure 1.64(a) shows a simple component drawn in isometric projection. Figure 1.64(b) shows the same component as an orthographic drawing. This time we make no attempt to represent the component pictorially. Each view of each face is drawn out separately either full size or to the same scale. What is important is how we position the various views as this determines how we 'read' the drawing.

Take a careful look at Figure 1.64 and note the following features.

Elevation
This is the main view from which all the other views are positioned. You look directly at the side of the component and draw what you see.

Plan
To draw this, you look directly down on the top of the component and draw what you see below the elevation (this is called a *plan view*).

End view
This is sometimes called an *end elevation*. To draw this you look directly at the end of the component and draw what you see at the opposite end of the elevation. There may be two end views, one at each end of the elevation, or there may be only one end view if this is all that is required to completely depict the component. Figure 1.64 requires only one end view. When there is only one end view this can be placed at either end of the elevation depending upon which position gives the greater clarity and ease of interpretation.

> **Key point**
>
> Engineers use two orthographic drawing techniques called *first-angle projection* (English projection) and *third-angle projection* (American projection). Using these techniques, designers are able to draw three different views of a part that are referred to as the *elevation, end view* and *plan view*.

Whichever end is chosen the rules for drawing this view must be obeyed.

The *construction lines* shown in Figure 1.64 can be removed when the drawing is complete leaving the completed first-angle projection as shown in Figure 1.65.

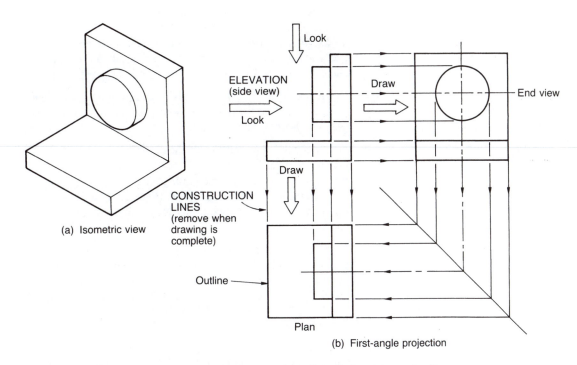

Figure 1.64 *An isometric view and its corresponding first-angle projection*

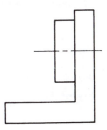
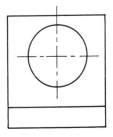

FIRST–ANGLE PROJECTION

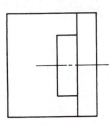

Figure 1.65 *Completed first-angle projection*

How to construct a first-angle projection

✓ You will need an A3 or A4 sketch pad or a sheet of drawing or sketching paper, an HB pencil or a drawing pen and an eraser. Alternatively, you will need access to a CAD system.

✓ Start by deciding how much space you will need for your first-angle projection (remember that you will need three views – not just one!). Also decide on what scale you will use, and how your drawing will be positioned on the drawing sheet.

✓ Construct the elevation. Remember that this is the main view from which all of the other views will be constructed so position the elevation with care!

✓ Now draw the plan view *below* the elevation. Do this by constructing feint construction lines (if using manual drawing techniques) and then lining them in more heavily to indicate the shape of the plan view. If you find this hard to do just follow the corresponding 'look' and 'draw' arrows shown in Figure 1.64.

✓ Next draw the end view *to the right* of the elevation. Once again, do this by drawing feint construction lines (if using manual drawing techniques) and then lining them in more heavily to indicate the shape of the end view. As before, if you find this difficult just follow the corresponding 'look' and 'draw' arrows shown in Figure 1.64.

✓ Complete the drawing showing all features and add labels and dimensions as required.

✓ Any lines that you are unhappy with (including construction lines) can be erased and, if necessary, redrawn.

✓ Complete the title block.

In order to make your first-angle drawing more useful you can

✓ Add a note saying which projection was used for the drawing.

Activity 1.25

Figure 1.66 shows a component drawn in isometric view. Redraw this component using first-angle projection and include dimensions. You should use either manual or CAD drawing techniques as instructed by your teacher or tutor. In either case you should make use of one of your blank drawing sheet templates. Complete the title block and add the finished drawing to your PDP.

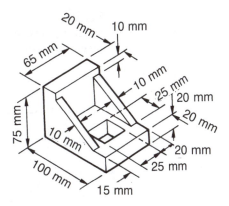

Figure 1.66 *See Activity 1.25*

Activity 1.26

Figure 1.67 shows a component drawn in cabinet oblique view. Redraw this component using first-angle projection and include dimensions. You should use either manual or CAD drawing techniques as instructed by your teacher or tutor. In either case you should make use of one of your blank drawing sheet templates. Complete the title block and add the finished drawing to your PDP.

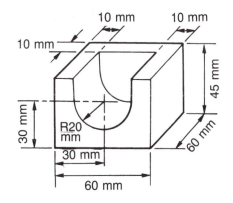

Figure 1.67 *See Activity 1.26*

1.4.5 Third-angle projection

Third-angle projection is an alternative to first-angle projection. Figure 1.68 shows the same component as that used in Figure 1.64 but this time we have drawn it in third-angle projection.

Take a careful look at Figure 1.68 and note the following features.

Elevation

Again we have started with the elevation or side view of the component and, as you can see, there is no difference.

Plan

Again we look down on top of the component to see what the plan view looks like. However, this time, we draw the plan view *above* the

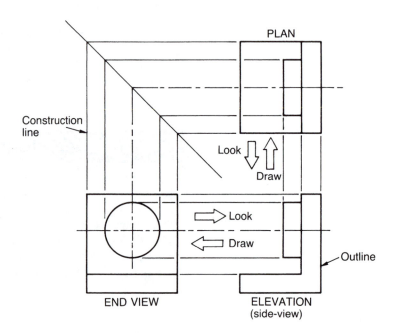

Figure 1.68 *Third-angle projection*

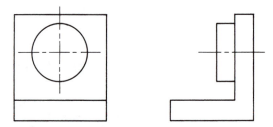

Figure 1.69 *Completed third-angle projection*

Test your knowledge 1.30

Explain, with the aid of a sketch, the difference between first- and third-angle projection.

elevation. That is, in third-angle projection we draw all the views from where we look.

End view

Note how the position of the end view is reversed compared with first-angle projection. This is because, like the plan view, we draw the end views at the same end from which we look at the component.

Once again, the *construction lines* shown in Figure 1.68 can be removed when the drawing is complete leaving the completed third-angle projection as shown in Figure 1.69.

How to construct a third-angle projection

✓ You will need an A3 or A4 sketch pad or a sheet of drawing or sketching paper, an HB pencil or a drawing pen and an eraser. Alternatively, you will need access to a CAD system.

✓ Start by deciding how much space you will need for your third-angle projection (remember that you will need three views – not just one!). Also decide on what scale you will use, and how your drawing will be positioned on the drawing sheet.

✓ Construct the elevation. Remember that this is the main view from which all of the other views will be constructed so position the elevation with care!

✓ Now draw the plan view *above* the elevation. Do this by constructing feint construction lines (if using manual drawing techniques) and then lining them in more heavily to indicate the shape of the plan view. If you find this hard to do just follow the corresponding 'look' and 'draw' arrows shown in Figure 1.68.

✓ Next draw the end view *to the left* of the elevation. Once again, do this by drawing feint construction lines (if using manual drawing techniques) and then lining them in more heavily to indicate the shape of the end view. As before, if you find this difficult just follow the corresponding 'look' and 'draw' arrows shown in Figure 1.68.

✓ Complete the drawing showing all features and add labels and dimensions as required.

✓ Any lines that you are unhappy with (including construction lines) can be erased and, if necessary, redrawn.

✓ Complete the title block.

In order to make your third-angle drawing more useful you can

✓ Add a note saying which projection was used for the drawing.

Activity 1.27

Figure 1.70 shows a component drawn in cabinet oblique view. Redraw this component using third-angle projection and include dimensions. You should use either manual or CAD drawing techniques as instructed by your teacher or tutor. In either case you should make use of one of your blank drawing sheet templates. Complete the title block and add the finished drawing to your PDP.

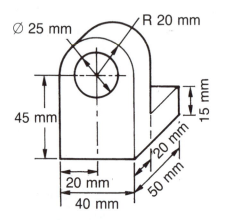

Figure 1.70 *See Activity 1.27*

1.4.6 Fluid power schematic diagrams

Fluid power schematics cover both pneumatic and hydraulic circuits. The symbols that we shall use do not illustrate the physical make-up, construction or shape of the components. Neither are the symbols to scale or orientated in any particular position. They are only intended to show the 'function' of the component they portray, the connections and the fluid flow path.

Complete symbols are made up from one or more basic symbols and from one or more functional symbols. Examples of some basic symbols are shown in Figure 1.71 and some functional symbols are shown in Figure 1.72.

Let us now see how we can combine some of these basic and functional symbols to produce a complete symbol representing a component. For example let us start with a motor. The complete symbol is shown in Figure 1.73.

The large circle indicates that we have an energy conversion unit such as a motor or pump. Notice that the fluid flow is into the device and that it is pneumatic. The direction of the arrowhead indicates the direction of flow. The fact that the arrowhead is clear (open) indicates that the fluid is air. Therefore the device must be a motor. If it were a pump the fluid flow would be *out* of the circle. The single line at the

Description	Symbol
Flow lines	
Continuous: Working line return line feed line	————
Long dashes: Pilot control lines	— — — —
Short dashes Drawn lines	- - - - - - -
Long chain enclosure line	— - - —
Flow line connections	
Mechanical link, roller, etc	O
Semi-rotary actuator	D
As a rule, control valves (valve) except for non-return valves	
Conditioning apparatus (filter, separator, lubricator, heat exchanger)	◇

Description	Symbol
Spring	⋀⋀⋁
Restriction: affected by viscosity	
unaffected by viscosity	
As a rule, energy conversion units (pump, com-pressor motor)	◯
Measuring instruments	◯
Non-return valve, rotary connection, etc	O

Figure 1.71 *Basic symbols used in fluid power diagrams*

The direction of flow and the nature of the fluid:	
hydraulic flow	▼
pneumatic flow or exhaust to atmosphere	▽
Indication of:	
direction	
direction of rotation	
path and direction of flow through valves	
As a general rule, the line perpendicular to the head of the arrow indicates that when the arrow moves the interior path always remains connected to the corresponding exterior path	
Indication of the possibility of a regulation or of a progressive variability	

Figure 1.72 *Functional symbols used in fluid power diagrams*

Figure 1.73 *Basic symbol for a motor*

Figure 1.74 *Energy converter symbol*

bottom of the circle is the outlet (exhaust) from the motor and the double line is the mechanical output from the motor. Now let us analyse the symbol shown in Figure 1.74:

- The circle tells us that it is an energy conversion unit
- The arrowheads show that the flow is from the unit so it must be a pump
- The arrowheads are solid so it must be a hydraulic pump
- The arrowheads point in opposite directions so the pump can deliver the hydraulic fluid in either direction depending upon its direction of rotation
- The arrow slanting across the pump is the variability symbol, so the pump has variable displacement
- The double lines indicate the mechanical input to the pump from some engine or motor.

Summing up, we have a variable displacement, hydraulic pump that is bi-directional.

Directional control valves (DCVs) are used to open or close the flow lines in a system. Control valve symbols are always drawn in square boxes or groups of square boxes to form a rectangle. This is how you recognise them. Each box indicates a discrete position for the control valve. Flow paths through a valve are known as *ways*. Thus a four-way valve has four flow paths through the valve. This will

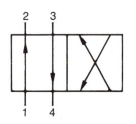

Figure 1.75 *4/2 directional control valve*

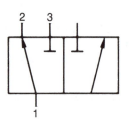

Figure 1.76 *See Test your knowledge 1.32*

be the same as the number of connections. We can, therefore, use a number code to describe the function of a valve. Figure 1.75 shows a 4/2 DCV. This valve has four flow paths, ports or connections and two positions. The two boxes indicate the two positions. The appropriate box is shunted from side to side so that, in your imagination, the internal flow paths line up with the connections. The lines that extend outside the perimeters of the boxes show connections.

As drawn, the fluid can flow into port 1 and out of port 2. Fluid can also flow into port 3 and out of port 4. In the second position, the fluid flows into port 3 and out of port 1. Fluid can also flow into port 4 and out of port 2.

Before we look at other examples of DCVs, let us see how we can control the positions of a valve. There are several basic methods of control, these are:

- Manual control of the valve position
- Mechanical control of the valve position
- Electromagnetic control of the valve position
- Pressure control of the valve positions (direct and indirect)
- Combined control methods.

The various methods that we use to control a valve are shown in Figure 1.77. With simple electrical or pressure control, it is only possible to move the valve to one, two or three discrete positions. The valve *spool* may be located in such positions by a spring-loaded *detent*.

Combinations of these control methods are possible. For example a single solenoid with spring return for a two-position valve. Let us now look at some further DCVs:

- Figure 1.78(a) shows a 4/2 DCV controlled by a single solenoid with a spring return.

- Figure 1.78(b) shows a 4/3 DCV. That is, a DCV with four ports (connections) and three positions. A lever operates it manually with spring return to the centre. The service ports are isolated in the centre position. An application of this valve will be shown later.

- Figure 1.78(c) shows a 4/2 DCV controlled by pneumatic pressure by means of a pilot valve. A single solenoid and a return spring actuate the pilot valve.

A *linear actuator* is a device for converting fluid pressure into a mechanical force capable of doing useful work and combining this force with limited linear movement. Put more simply, a piston in a cylinder. The symbols for linear actuators (also known as 'jacks' and 'rams') are simple to understand and some examples are shown in Figure 1.80:

- Figure 1.80(a) shows a single-ended, double-acting actuator. That is, the piston is connected by a piston rod to some external mechanism through one end of the cylinder only. It is double acting because fluid pressure can be applied to either side of the piston.

- Figure 1.80(b) shows a single-ended, single-acting actuator with spring return. Here the fluid pressure is applied only to

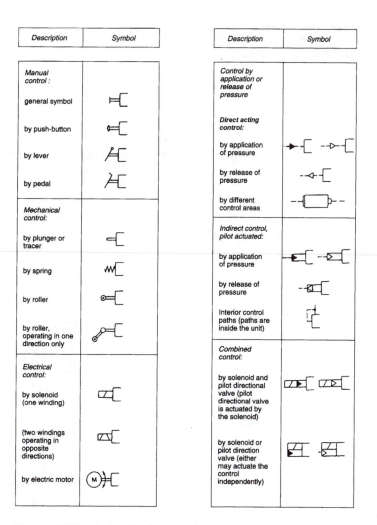

Description	Symbol
Manual control : general symbol	
by push-button	
by lever	
by pedal	
Mechanical control: by plunger or tracer	
by spring	
by roller	
by roller, operating in one direction only	
Electrical control: by solenoid (one winding)	
(two windings operating in opposite directions)	
by electric motor	

Description	Symbol
Control by application or release of pressure *Direct acting control:* by application of pressure	
by release of pressure	
by different control areas	
Indirect control, pilot actuated: by application of pressure	
by release of pressure	
Interior control paths (paths are inside the unit)	
Combined control: by solenoid and pilot directional valve (pilot directional valve is actuated by the solenoid)	
by solenoid or pilot direction valve (either may actuate the control independently)	

Figure 1.77 *Methods of control*

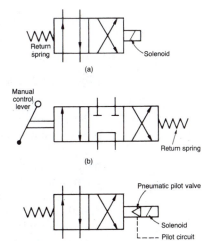

Figure 1.78 *Various types of DCV*

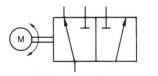

Figure 1.79 *See Test your knowledge 1.33*

one side of the piston. Note the pneumatic exhaust to atmosphere so that the air behind the piston will not cause a fluid lock.

- Figure 1.80(c) shows a single-ended, single-acting actuator, with double variable cushion damping. The cushion damping prevents the piston impacting on the ends of the cylinder and causing damage.

- Figure 1.80(d) shows a double-ended, double-acting actuator fitted with single, fixed cushion damping.

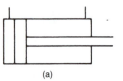

(a)

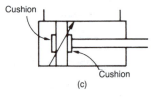

(b)

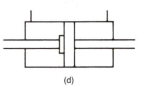

(c)

(d)

Figure 1.80 *Various types of linear actuator*

We are now in a position to use the previous component symbols to produce some simple fluid power circuits.

Figure 1.81 shows a single-ended, double-acting actuator controlled by a 4/3 tandem centre, manually operated DCV. Note that in the neutral position both sides of the actuator piston are blocked off, forming a hydraulic lock. In this position the pump flow is being returned directly to the tank. Note the tank symbol. This system is being supplied by a single direction fixed displacement hydraulic pump.

Figure 1.82 shows a simple *pneumatic hoist* capable of raising a load. The circuit uses two 2-port manually operated push-button valves connected to a single-ended, single-acting actuator. Supply

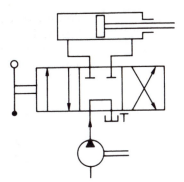

Figure 1.81 *Actuator controlled by a DCV*

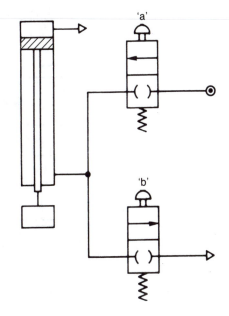

Figure 1.82 *A simple pneumatic hoist*

pressure is indicated by the circular symbol with a black dot in its centre. Valve 'b' has a threaded exhaust port indicated by the extended arrow. When valve 'a' is operated, compressed air from the airline is admitted to the underside of the piston in the cylinder. This causes the piston to rise and to raise the load. Any air above the piston is exhausted to the atmosphere through the threaded exhaust port at the top of the cylinder. Again a long arrow indicates this. When valve 'b' is operated, it connects the cylinder to the exhaust and the actuator is vented to the atmosphere. The load is lowered by gravity.

Both of these circuits are functional, but they do not have protection against over-pressurisation, neither do they have any other safety devices fitted. Therefore, we need to increase our vocabulary of components before we can design a safe, practical circuit. We will now consider the function and use of pressure and flow control valves.

Figure 1.83 shows an example of a pressure relief (safety) valve. In Figure 1.83(a) the valve is being used in a hydraulic circuit.

Pressure is controlled by opening the exhaust port to the reservoir tank against an opposing force such as a spring. In Figure 1.83(b) the valve is being used in a pneumatic circuit so it exhausts to the atmosphere.

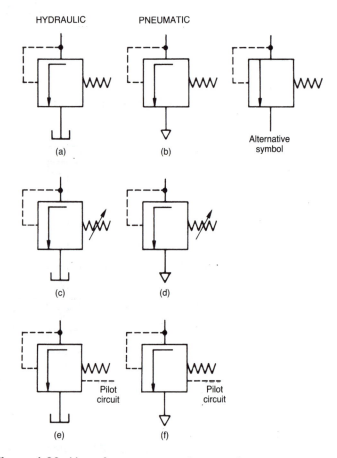

Figure 1.83 *Use of a pressure release valve*

Figures 1.83(c) and (d) show the same valves except that this time the relief pressure is variable, as indicated by the arrow drawn across the spring. If the relief valve setting is used to control the normal system pressure as well as acting as an emergency safety valve, the adjustment mechanism for the valve must be designed so that the maximum safe working pressure for the circuit cannot be exceeded.

Figures 1.83(e) and (f) show the same valves with the addition of pilot control. This time the pressure at the inlet port is not only limited by the spring but also by the pressure of the pilot circuit superimposed on the spring. The spring offers a minimum pressure setting and this can be increased by increasing the pilot circuit pressure up to some pre-determined safe maximum. Sometimes the spring is omitted and only pilot pressure is used to control the valve.

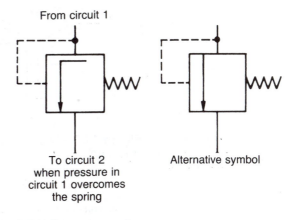

From circuit 1

To circuit 2
when pressure in
circuit 1 overcomes
the spring

Alternative symbol

Figure 1.84 *Sequence valve*

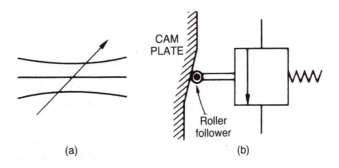

CAM
PLATE

Roller
follower

(a)

(b)

Figure 1.85 *Flow control valve*

Sequence valves are closely related to relief valves in both design and function and are represented by very similar symbols. They permit the hydraulic fluid to flow into a sub-circuit, instead of back to the reservoir, when the main circuit pressure reaches the setting of the sequence valve. You can see that Figure 1.84 is very similar to a pressure relief valve (PRV) except that, when it opens, the fluid is directed to the next circuit in the sequence instead of being exhausted to the reservoir tank or allowed to escape to the atmosphere.

Flow control valves, as their name implies, are used in systems to control the rate of flow of fluid from one part of the system to another. The simplest valve is merely a fixed restrictor. For operational reasons this type of flow control valve is inefficient, so the restriction is made variable as shown in Figure 1.85(a). This is a throttling valve. The full symbol is shown in Figure 1.85(b). In this example the valve setting is being adjusted mechanically. The valve rod ends in a roller follower in contact with a cam plate.

Sometimes it is necessary to ensure that the variation in inlet pressure to the valve does not affect the flow rate from the valve. Under these circumstances we use a pressure compensated flow control valve (PCFCV). The symbol for this type of valve is shown in Figure 1.86. This symbol suggests that the valve is a combination of a variable restrictor and a pilot operated relief valve. The enclosing box is drawn using a long-chain line. This signifies that the components making up the valve are assembled as a single unit.

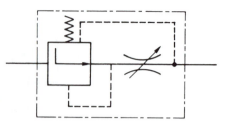

Figure 1.86 *Pressure compensated flow control valve (PCFCV)*

The *non-return valve* (NRV), or *check valve* as it is sometimes known, is a special type of DCV. It only allows the fluid to flow in one direction and it blocks the flow in the reverse direction. These valves may be operated directly or by a pilot circuit. Some examples are shown in Figure 1.87:

- Figure 1.87(a) shows a valve that opens (is free) when the inlet pressure is higher than the outlet pressure (back pressure).

- Figure 1.87(b) shows a spring-loaded valve that only opens when the inlet pressure can overcome the combined effects of the outlet pressure and the force exerted by the spring.

- Figure 1.87(c) shows a pilot-controlled NRV. It only opens if the inlet pressure is greater than the outlet pressure. However, these pressures can be augmented by the pilot circuit pressure.

- The pilot pressure is applied to the inlet side of the NRV. We now have the combined pressures of the main (primary) circuit and the pilot circuit acting against the outlet pressure. This enables the valve to open at a lower main circuit pressure than would normally be possible.

- The pilot pressure is applied to the outlet side of the NRV. This assists the outlet or back pressure in holding the valve closed. Therefore it requires a greater main circuit pressure to open the valve. By adjusting the pilot pressure in these two examples we can control the circumstances under which the NRV opens.

- Figure 1.87(d) shows a valve that allows normal full flow in the forward direction but restricted flow in the reverse direction. The valves previously discussed did not allow any flow in the reverse direction.

- Figure 1.87(e) shows a simple shuttle valve. As its name implies, the valve is able to shuttle backwards and forwards. There are two inlet ports and one outlet port. Imagine that inlet port A has the higher pressure. This pressure overcomes the inlet pressure at B and moves the shuttle valve to the right. The valve closes inlet port B and connects inlet port A to the outlet port. If the pressure at inlet port B rises, or that at A falls, the shuttle will move back to the left. This will close inlet port A and connect inlet port B to the outlet. Thus, the inlet port with the higher pressure is automatically connected to the outlet port.

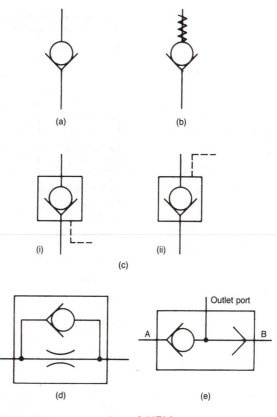

Figure 1.87 *Some examples of NRV*

The working fluid, be it oil or air, has to operate in a variety of environments and it can become overheated and/or contaminated. As its name implies, conditioning equipment is used to maintain the fluid in its most efficient operating condition. A selection of conditioning equipment symbols is shown in Figure 1.88. Note that all conditioning device symbols are diamond shaped.

Filters and *strainers* have the same symbol. They are normally identified within the system by their position. The filter element (dotted line) is always positioned at 90° to the fluid path.

Water traps are easily distinguished from filters since they have a drain connection and an indication of trapped water. Water traps are particularly important in pneumatic systems because of the humidity of the air being compressed.

Lubricators are particularly important in pneumatic systems. Hydraulic systems using oil are self-lubricating. Pneumatic systems use air, which has no lubricating properties so oil, in the form of a mist, has to be added to the compressed air line.

Heat exchangers can be either heaters or coolers. If the hydraulic oil becomes too cool it becomes thicker (more viscous) and the system becomes sluggish. If the oil becomes too hot it will become too thin (less viscous) and not function properly. The direction of the arrows in the symbol indicates whether heat energy is taken from the fluid (cooler) or given to the fluid (heater). Notice that the cooler can show the flow lines of the coolant.

Filters, water traps, lubricators and miscellaneous apparatus

Description	Symbol
Filter or strainer	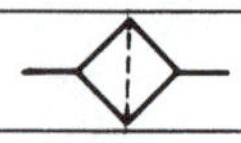
Water trap: with manual control	
automatically drained	
Filter with water trap: with manual control	
automatically drained	
Air dryer	
Lubricator	
Conditioning unit detailed symbol	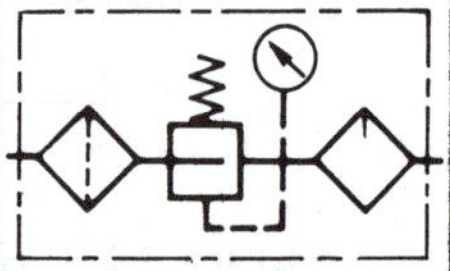
simplified symbol	

Heat exchangers

Description	Symbol
Temperature controller (arrows indicate that heat may be either introduced or dissipated)	
Cooler (arrows indicate the extraction of heat) without representation of the flow lines of the coolant	
with representation of the flow lines of the coolant	
Heater (arrows indicate the introduction of heat)	

Figure 1.88 *Symbols for conditioning devices*

There is one final matter to be considered before you can try your hand at designing a circuit, and that is the pipework circuit to connect the various components together. The correct way of representing pipelines is shown in Figure 1.90:

- Figure 1.90(a) shows pipelines that are crossing each other but are not connected.

- Figure 1.90(b) shows three pipes connected at a junction. The junction (connection) is indicated by the solid circle (or large dot, if you prefer).

- Figure 1.90(c) shows four pipes connected at a junction. On no account can the connection be drawn as shown in Figure 1.90(d). This is because there is always a chance of the ink running where lines cross on a drawing. The resulting 'blob' could then be misinterpreted as a connection symbol with disastrous results!

This concludes our work on fluid power schematics and the components that are used in them. Next we will be looking at electrical and electronic schematic diagrams.

Test your knowledge 1.34

Figure 1.89 shows a selection of fluid circuit symbols. Name the symbols and briefly explain what they do.

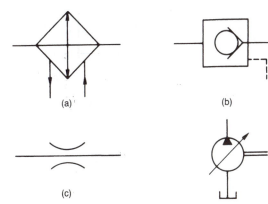
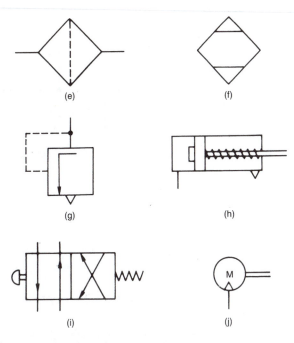

Figure 1.89 *See Test your knowledge 1.34*

Activity 1.28

Figure 1.91 shows the general principles for the hydraulic drive to the ram of a shaping machine. The ram is moved backwards and forwards by a double-acting single-ended hydraulic actuator. The drawing was made many years ago and it uses outdated symbols. Use CAD or manual drawing techniques (as instructed by your teacher or tutor) to draw a schematic hydraulic diagram for the machine using current symbols and practices (as set out in BS PP 7307). Present your work in the form of a printed diagram and include it in your PDP.

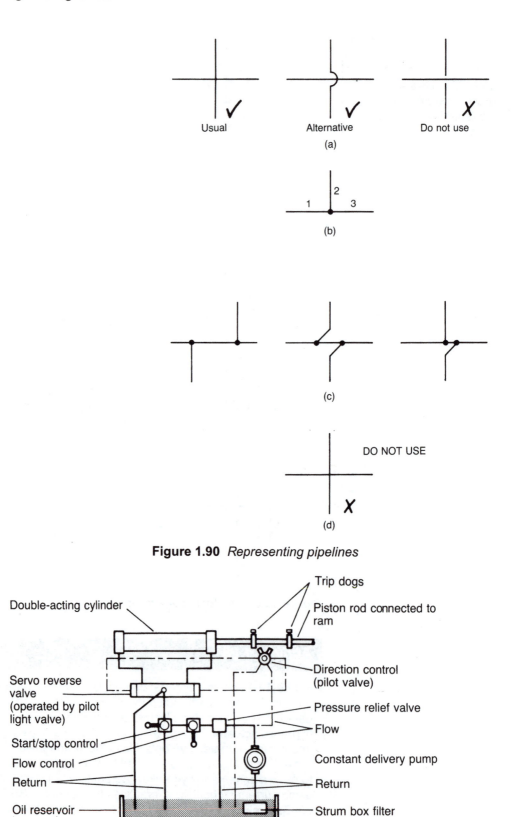

Figure 1.90 *Representing pipelines*

Figure 1.91 *See Activity 1.28*

1.4.7 Electrical and electronic circuit schematic diagrams

Like pneumatic and hydraulic circuits, electrical and electronic circuits are also drawn using schematic symbols to represent the various components. The full range of symbols and their use can be found in BS 3939 but you will find information suitable for schools and colleges in BS PP 7307. Figure 1.92 shows a selection of symbols that will be used in the following examples:

- A *cell* is a source of direct current (d.c.) electrical energy. Primary cells have a nominal potential of 1.5 V each. They cannot be recharged and are disposable. Secondary cells are rechargeable. Lead–acid cells have a nominal potential of 2 V and nickel cadmium (NiCd) cells have a nominal potential of 1.2 V. Cells are often connected in series to form a battery.

- *Batteries* consist of a number of cells connected in series to increase the overall potential. A 12 V car battery consists of six lead–acid secondary cells of 2 V each.

- *Fuses* protect the circuit in which they are connected from excess current flow. This can result from a fault in the circuit, from a fault in an appliance connected to the circuit or from too many appliances being connected to the same circuit. The current flowing in the circuit tends to heat up the fuse wire. When the current reaches some pre-determined value the fuse wire melts and breaks the circuit so the current can no longer flow. Without a fuse the circuit wiring could overheat and cause a fire.

- *Resistors* are used to control the magnitude of the current flowing in a circuit. The resistance value of the resistor may be fixed or it may be variable. Variable resistors may be preset or they may be adjustable by the user. The electric current does work in flowing through the resistor and this heats up the resistor. The resistor must be chosen so that it can withstand this heating effect and sited so that it has adequate ventilation.

- *Capacitors*, like resistors, may be fixed in value or they may be preset or variable. Capacitors store electrical energy but, unlike secondary cells, they may be charged or discharged almost instantaneously. The stored charge is much smaller than the charge stored by a secondary cell. Large value capacitors are used to smooth the residual ripple from the rectifier in a power pack. Medium value capacitors are used for coupling and decoupling the stages of audio frequency amplifiers. Small value capacitors are used for coupling and decoupling radio frequency signals and they are also used in tuned (resonant) circuits.

- *Inductors* act like electrical 'flywheels'. They limit the build up of current in a circuit and try to keep the circuit running by putting energy back into it when the supply is turned off. They are used as current limiting devices in fluorescent lamp units. They are used as chokes in telecommunications equipment. They are also used together with capacitors to make up resonant (tuned) circuits in telecommunications equipment.

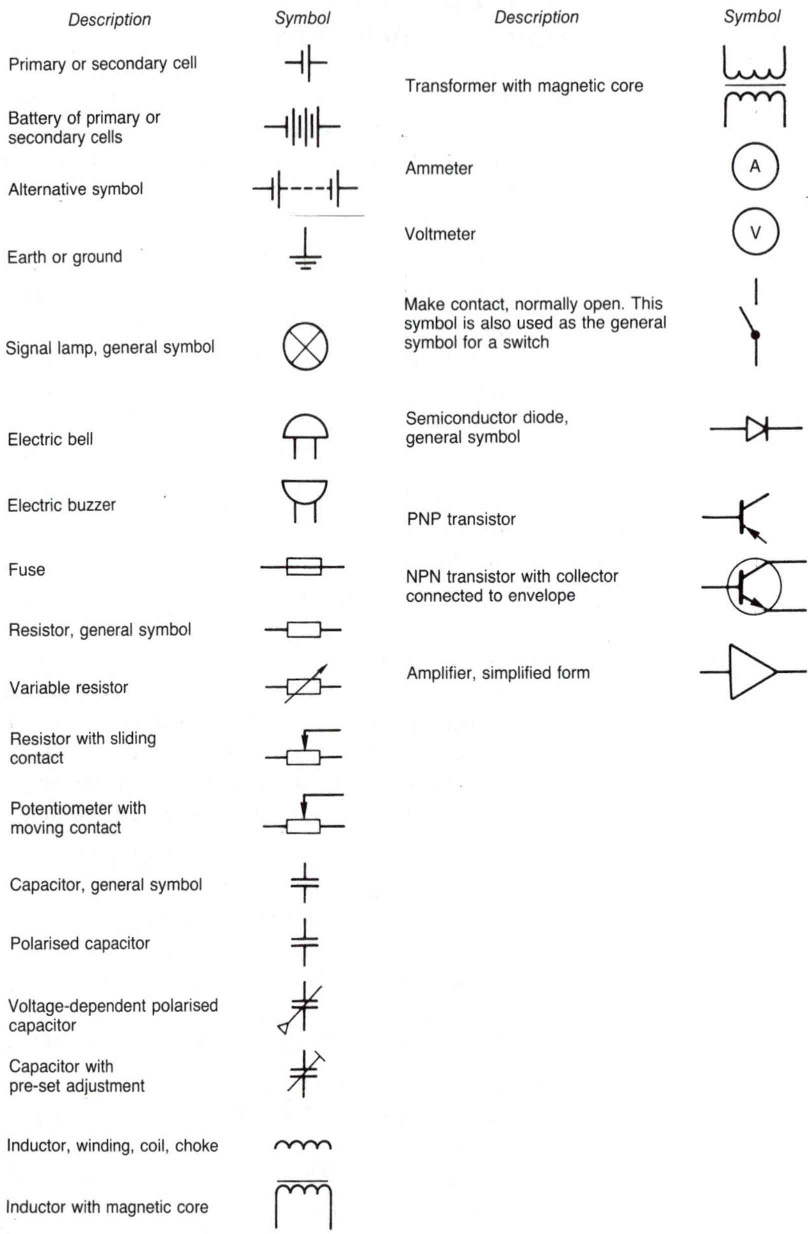

Description	Symbol	Description	Symbol
Primary or secondary cell		Transformer with magnetic core	
Battery of primary or secondary cells			
Alternative symbol		Ammeter	
Earth or ground		Voltmeter	
Signal lamp, general symbol		Make contact, normally open. This symbol is also used as the general symbol for a switch	
Electric bell		Semiconductor diode, general symbol	
Electric buzzer		PNP transistor	
Fuse		NPN transistor with collector connected to envelope	
Resistor, general symbol		Amplifier, simplified form	
Variable resistor			
Resistor with sliding contact			
Potentiometer with moving contact			
Capacitor, general symbol			
Polarised capacitor			
Voltage-dependent polarised capacitor			
Capacitor with pre-set adjustment			
Inductor, winding, coil, choke			
Inductor with magnetic core			

Figure 1.92 *Electronic symbols*

- *Transformers* are used to raise or lower the voltage of alternating currents (a.c.). Inductors and transformers cannot be used in d.c. circuits. You cannot get something for nothing, so if you increase the voltage you decrease the current accordingly so that (neglecting losses), $V \times I = k$, where k is a constant for the primary and secondary circuits of any given transformer.

- *Ammeters* measure the current flowing in a circuit. They are always wired in series with the circuit so that the current being measured can flow through the meter.

- *Voltmeters* measure the potential difference (voltage) between two points in a circuit. To do this they are always wired in parallel across that part of the circuit where the potential is to be measured.

- *Switches* are used to control the flow of current in a circuit. They can only open or close the circuit. So the current either flows or it does not.

- *Diodes* are like the NRVs in hydraulic circuits. They allow the current to flow in one direction only as indicated by the arrowhead of the symbol. They are used to rectify a.c. and convert it into d.c.

- *Transistors* are used in high-speed switching circuits and to magnify radio and audio frequency signals.

- *Integrated circuits* consist of all the components necessary to produce amplifiers, oscillators, central processor units, computer memories and a host of other devices fabricated onto a single slice of silicon; each chip being housed in a single compact package.

Let us look at some examples of schematic circuit diagrams using these symbols. All electric circuits consist of:

- A source of electrical energy (e.g. a battery or a generator).

- A means of controlling the flow of electric current (e.g. a switch or a variable resistor).

- An appliance to convert the electrical energy into useful work (e.g. a heater, a lamp or a motor).

- Except for low-power battery operated circuits, an over-current protection device (fuse or circuit breaker).

- Conductors (wires) to connect these various circuit elements together. Note that the rules for drawing conductors that are connected and conductors that are crossing but not connected are the same as for drawing pipework as previously described in Figure 1.90.

Figure 1.93 shows a very simple circuit that satisfies the above requirements. In Figure 1.93(a) the switch is 'closed' therefore the circuit as a whole is also a closed loop. This enables the electrons that make up the electric current to flow from the source of electrical

> **Key point**
>
> Electrical and electronic circuit schematics use standard symbols to represent the individual component parts. These symbols are joined by lines representing the wires (and other conductors such as printed circuit board tracks) that connect the parts together.

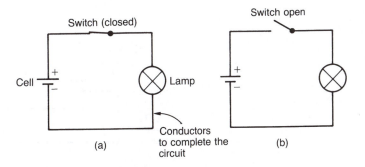

Figure 1.93 *A simple electronic circuit*

energy through the appliance (lamp) and back to the source of energy ready to circulate again. Rather like the fluid in our earlier hydraulic circuits. In Figure 1.93(b) the switch is 'open' and the circuit is no longer a closed loop. The circuit is broken. The electrons can no longer circulate. The circuit ceases to function. We normally draw our circuits with the switches in the 'open' position so that the circuit is not functioning and is therefore 'safe'.

Figure 1.94 shows a simple battery operated circuit for determining the resistance of a fixed value resistor. The resistance value is obtained by substituting the values of current and potential into the formula, $R = V/I$. The current in amperes is read from the ammeter and the potential in volts is read from the voltmeter. Note that the ammeter is wired in series with the resistor so that the current can flow through it. The voltmeter is wired in parallel with the resistor so that the potential can be read across it. This is always the way these instruments are connected.

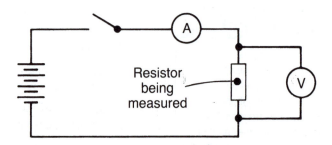

Figure 1.94 *Circuit for determining resistance*

Figure 1.95 shows a circuit for operating the light over the stairs in a house. The light can be operated either by the switch at the bottom of the stairs or by the switch at the top of the stairs. Can you work out how this is achieved? The switches are of a type called *two-way single-pole*. The circuit is connected to the mains supply. It is protected by a fuse in the *consumer unit*. This unit contains the main switch and all the fuses for the house and is situated adjacent to the supply company's meter and main fuse.

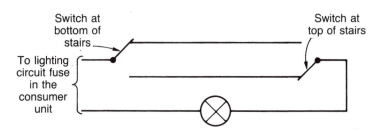

Figure 1.95 *Two-way lighting switch*

Figure 1.96 shows a two-stage transistorised amplifier. It also shows a suitable power supply. The components used in the circuit are listed below:

Component	Description
R_1–R_9	Fixed resistors
VR_1	Variable resistor
C_1–C_9	Capacitors
D_1–D_4	Diodes
TR_1, TR_2	Transistors
T_1	Mains transformer
L_1	Inductor (choke)

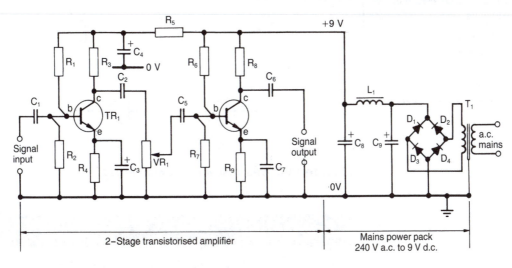

Figure 1.96 *A two-stage transistor amplifier*

Figure 1.97 shows a similar amplifier using a single integrated circuit *chip*. Such an amplifier would have the same performance but fewer components are required. Therefore it is cheaper and quicker to make.

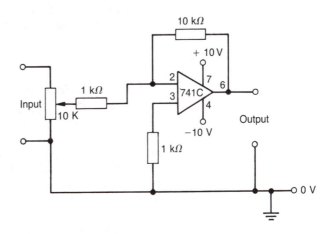

Figure 1.97 *A single-chip amplifier circuit*

Figure 1.98 shows a selection of electrical and electronic symbols. Name the symbol and briefly explain what they do.

Sketch standard symbols for the following electrical/electronic components:

(a) a fixed resistor
(b) a capacitor
(c) an inductor
(d) a diode.

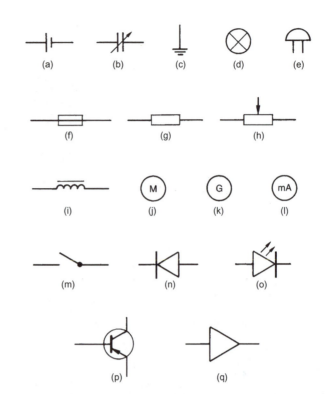

Figure 1.98 *See Test your knowledge 1.35*

Figure 1.99 shows an electronic circuit. You should use either manual or CAD drawing techniques as instructed by your teacher or tutor to make a formal engineering drawing of this circuit using one of your blank drawing sheet templates. Complete the title block and add the finished drawing to your PDP. You should also draw up a component list that numbers and names each of the components and includes values where given.

1.4.8 Exploded views

The final type of specialised drawing that we shall be looking at is called an *exploded view*. Quite simply, an exploded view is a pictorial representation of a product that is taken apart. By drawing the individual component parts separately but in approximately the same physical relationship as when assembled, you can gain a very good idea of how something is put together. Exploded views are very useful when a product has to be serviced or maintained. A service or maintenance engineer has only to take a look at an exploded diagram to see how the various parts fit together.

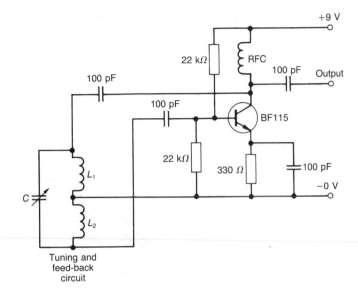

Figure 1.99 *See Activity 1.29*

Typical exploded diagrams are shown in Figures 1.100 and 1.101. Figure 1.100 shows the exploded diagram of the tailstock assembly of a lathe whilst Figure 1.101 shows an exploded view of a hi-fi amplifier chassis. Note that, in both cases, part numbers are included on the diagram.

1.5 Presenting a design solution

The time will come when you will have to explain your final design solution to other people. Your presentation must:

- give reasons for your final choice that refer to the key features in the design brief and your design specification
- show details of your final design idea
- give an explanation of how your final design solution meets the client design brief
- respond to feedback, checking against the design criteria and suitability for the user, and modify your proposed solution, if necessary.

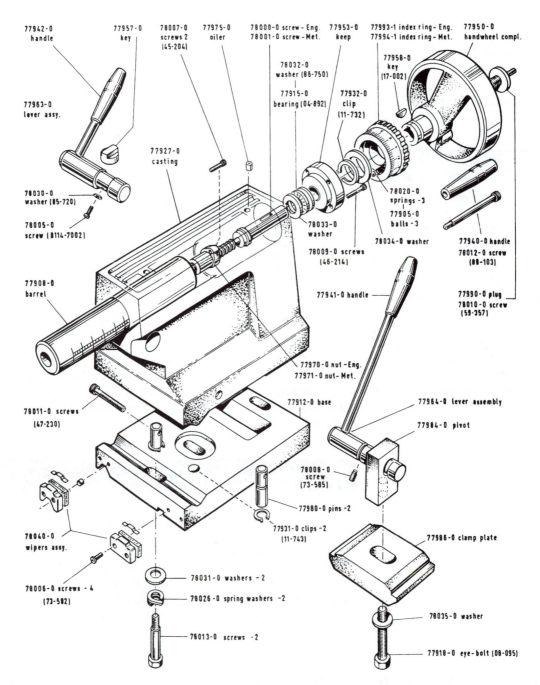

Figure 1.100 *An exploded view of the tailstock of a lathe*

When you work on a design project it is a good idea to keep a *design folder* containing all of the notes, sketches and drawings that you use at each phase of the design process. A design folder should contain notes, sketches and drawings showing what was done, why it was done, when it was done, and who did it. The design folder will later become invaluable when it comes to presenting your design solution to other people!

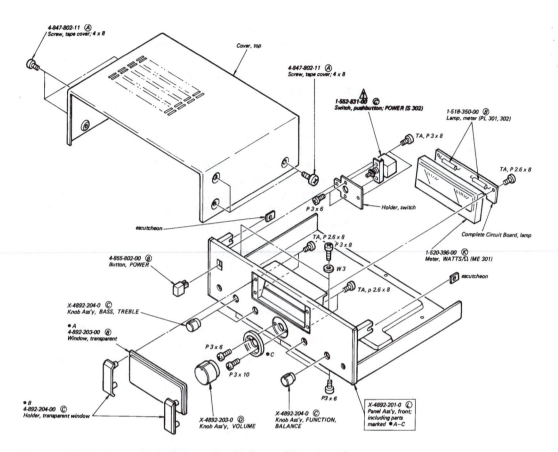

Figure 1.101 *An exploded view of a hi-fi amplifier chassis*

A design folder will typically contain the following items (see Figure 1.102).

A description of the problem
The problem is the task set to you by a client or you have given yourself. This section should be quite brief and will normally just be a few sentences that describe the problem. However, in some cases you may wish to add some sketches or drawings to enhance or clarify your description.

A statement of the design brief
The design brief states what needs to be done to solve the problem that you have identified. It should be written from the perspective of the client. In other words, it should say what the client is looking for. You will normally wish to agree the design brief with your client so it is important to get the wording right! Once again, this section is usually just text but here again, there may be occasions when you might wish to use a drawing or a sketch to clarify and enhance your wording.

Results of the research and investigation
You will need to summarise the results of your research and investigation using charts, diagrams, tables and any other method that helps convey the results to your readers. Since charts and diagrams can usually be easily understood this is often the best way to represent the

outcome of your research and investigation. One important point is that you should always quote the source of any data that you use. You should also provide a copy of any questionnaire or poll that you used.

The investigation may also consider other aspects such as identifying primary and secondary markets for the product or service, existing products and service that may be similar or competing in the same market, materials and manufacturing constraints, ergonomics and aesthetics, issues relating to standards and Health and Safety.

A summary of the candidate solutions and the process that you went through to generate them

List each of the candidate solutions that you arrived at and explain how you arrived at them. Give brief details of any brainstorming session that you held or include a mind map if you used one.

The design criteria and the design specification

Include the design criteria that you established as well as the detailed design specification.

The candidate solutions

You should describe the design ideas for each of our candidate solutions and the process by which you arrived at the final design solution. If you used an evaluation matrix this should be included here. You should include details and sketches of the ideas that you rejected. Although these ideas may not be taken forward, they give an idea of

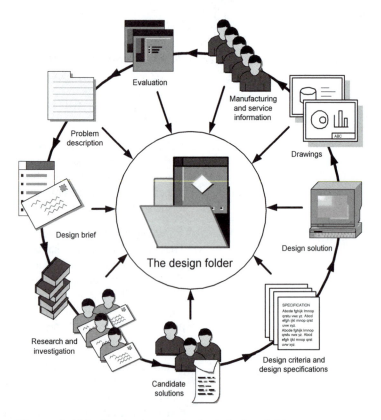

Figure 1.102 *Contents of a typical design folder*

the process that you went through and, at some time in the future, you or your client may want to return to them!

The final design solution

You will need to describe your final design solution in detail. To do this, you will need to use detailed sketches and drawings. These will help to convey your ideas to the client and they may also assist with marketing the product or service that you have designed.

Your working and presentation drawings

Your working drawings must be precise and drawn to scale. They must give specific information about dimensions and the materials that should be used. You may wish to include drawings of individual parts as well as details showing how your product should be assembled. Dimensions should be clearly indicated and all other relevant information should be included. One or more presentation drawings should be included to show what the finished product will look like. You should choose an appropriate drawing technique for this (perspective, isometric, oblique, 3D rendered CAD, etc.).

Information relating to manufacturing the product or supplying the service

You will need to provide details showing how you envisage that your product will be manufactured or how your service will be delivered. For example, you may need to supply a parts list, a component list or a cutting diagram. Much will depend on the individual product or service. Finally, you should provide a simple step-by-step explanation of what needs to be done to assemble or construct your designed product. A series of sketches might be useful here!

A final evaluation of your work

Your design folder will not be complete without a full evaluation of the work that you have done. You should also include comments and feedback that you received from your client as well as information on any modifications that were incorporated as the work went on. In many design projects your client will insist on holding a regular review meeting in order to discuss progress with the project. If your client does not ask for this, you might find it useful to organise a presentation of your work or provide a brief written report at various stages of the project.

You need to justify your final design solution by demonstrating that you have been able to comply with each of the essential design criteria as well as meeting the design specification that you agreed with your client. You also need to say how you complied with any relevant standards or legislation.

You need to look at your work with a critical eye and indicate where there are any particular strengths or weaknesses. You should also provide full details of any testing or measurement that you carried out. If your measurements did not confirm that you have met the design specification you need to suggest why this is and what should be done to ensure that the design specification is met.

Finally, you need to ask yourself whether your final design solution meets your client's needs and expectations. You should also comment on any improvements or modifications that could be made to improve the product or service or make it more cost effective.

How to organise a design folder

You need to ensure that your design folder includes sections that adequately cover each of the following aspects:

✓ A description of the problem.

✓ The design brief.

✓ Details of the research and investigation carried out.

✓ The ideas that you considered *and* a description of how they were generated.

✓ A brief summary of the candidate solutions.

✓ The design criteria and the design specification.

✓ Design ideas and the final design solution.

✓ A set of working and presentation drawings.

✓ Information relating to manufacturing the product or to supplying the service (as appropriate).

✓ A thorough final evaluation.

In order to make your design folder more useful you can

✓ Include relevant illustrations, sketches, line drawings and photographs.

✓ Organise your folder into numbered sections and subsections.

✓ Include a contents list on the first page and an index on the last page.

✓ Provide a full list of references and other information sources that you used in the various stages of the design project.

If you are required to give a formal presentation of your work this might involve either a verbal presentation supported by appropriate visual aids (e.g. a PowerPoint presentation or overhead projector transparencies) or a written report. In either case, your presentation should be delivered in a way that is appropriate to your audience. This will invariably mean that you should keep your presentation brief and to the point. At the same time, you should ensure that you have covered all the main points that make up your design solution. In any event, your presentation must be interesting and appropriately paced so that the attention of the audience does not wander. If you are delivering a verbal presentation it is also important to be a good listener and be able to respond to any questions or queries that are raised by your audience.

Fortunately, much of the material that you need to include in your verbal presentation will already exist in your design folder. All you need to do is to summarise it and present it in a way that your audience can quickly and easily grasp. In most cases, you will wish to supply your audience with a set of notes or printed *handouts*. These can be based on PowerPoint screens of overhead projector transparencies and can be augmented with sketches and presentation drawings, as appropriate.

Activity 1.31

Use presentation software (e.g. PowerPoint) to prepare a 5-min presentation to the rest of the class (using appropriate visual aids) on any one of the following topics:

- How to choose a digital camera
- How to connect to the Internet
- What to look for when purchasing a second-hand car.

You should prepare a set of brief printed notes summarising the key points for your audience. Also include printed copies of any screens or overhead projector transparencies that you use. At the end of your talk you should invite questions from your audience and provide appropriate answers. Add your presentation notes to your PDP.

The alternative to a formal verbal presentation is that of providing your client with a written technical report. Once again, this can be constructed from material in your design folder. Typical section headings that you might wish to include in your technical report might include the following.

Summary
A brief overview for busy readers who need to quickly find out what the report is about.

Introduction
This sets the context and background and provides a brief description of the problem that you have solved and may also include a statement of the design brief.

Main body
A comprehensive description of the design solution including how and why it was chosen together with details of the research and investigation that you carried out. You should also include (and comment on) the design criteria and the final design specification. Your design solution should be presented together with sketches and drawings.

Evaluation
A detailed evaluation of the work that you carried out including any problems that you encountered and how you solved them.

Recommendations
This section should provide information on how the design solution should be implemented (including, e.g. information on manufacture or assembly). You may also wish to include any modifications or changes that you would recommend.

Conclusions

You should end your report with a few concluding remarks about your design solution and how effective you think it is likely to be at solving your client's problem.

References

This section should provide readers with a list of sources for further information relating to the scientific principles or technology used, including (where appropriate) relevant standards and legislation. Finally, it is important to take care when you present your work in the form of a written report. To avoid confusion, the normal conventions of grammar and punctuation must be used. Words must be correctly spelt so do make use of a dictionary if you are uncertain about spelling. If you are using a word processing package use the spell-checker. Never use jargon terms and acronyms unless you are sure that the persons reading the message are as equally familiar with them as is the writer. Layout is important so use numbered sections and paragraphs and try to keep your sentences short and to the point!

How to write a technical report

Most of the material that you need for your technical report will already be in your design folder so it is important to ensure that it is up to date *before* you start!

✓ Describe the problem and state the design brief within your *Introduction*.

✓ Start by describing your *Design Solution*. Say how and why it was chosen. This should appear in a subsection of the *Main Body* of your report.

✓ Summarise the *Research and Investigation* that you carried out and include this as another subsection of the *Main Body* of your report.

✓ List the *Design Criteria* and the final *Design Specification* and include these as further subsections of the *Main Body* of your report. Also give brief details of the other ideas that you considered and say why they were rejected.

✓ Present your *Design Solution* using sketches, working and presentation drawings (as appropriate) and include these as further subsections in the *Main Body* of your report.

✓ Describe any problems that you encountered and how you overcame them. Identify any particular strengths and weaknesses of your solution. Include all of this within the *Evaluation* section.

✓ Suggest how your work could have been improved. Describe any modifications that could be made and give details of any enhancements that could be incorporated. This information should appear in the *Recommendations* section.

✓ Write a short *Conclusion* saying how closely the design solution has been able to satisfy the client's needs and say whether or not it is likely to provide an adequate solution to the design problem. Keep your conclusion short and to the point.

✓ Include a *References* section that will provide your readers with details of the sources of information that you used at various stages of the design project.

✓ Write a brief overview of your report and include this in the *Summary*. Note that your summary should say what the report is about *not* what the design solution is!

In order to make your technical report more useful you can

✓ Check that you have included all of the following main sections in your technical report: *Summary, Introduction, Main Body, Evaluation, Recommendations, Conclusions, References*. The *Main Body* should be further divided into sections headed *Design Solution, Research and Investigation, Design Criteria, Design Specification*, etc.

✓ Organise your folder into numbered sections and subsections.

✓ Include a contents list on the first page and an index on the last page.

✓ Include plenty of sketches, diagrams and drawings!

1.6 Revision problems

1. Describe, briefly, the stages in a typical design project and list them in the order in which they are carried out.

2. Explain the difference between an internal client and an external client.

3. Explain what a design brief is used for and what it should consist of.

4. Describe TWO ways in which research information can be obtained from the intended users of an engineered product or service.

5. The following data was obtained from potential users of a mobile car tune-up service:

Question: *What is your annual vehicle mileage?*

Responses: *Less than 5000 miles* 466 responses
5000 to 9999 miles 1259 responses
10,000 to 14,999 miles 871 responses
More than 15,000 miles 225 responses

Show how this information can be represented as:

(a) A bar chart. (b) A pie chart.

6. Describe TWO ways of generating ideas.

7. Explain, with the aid of a simple example, the use of an evaluation matrix.

8. A design brief involves manoeuvring a satellite into a precise orbit around the earth. The design solution involves firing one of several small rocket motors for an exact amount of time until the required position is achieved. Briefly explain the scientific principle that underpins the design solution.

9. List FIVE primary design needs that should appear within a design specification.

10. Explain, with the aid of a sketch, the relationship that exists between the most commonly used 'A' sizes of drawing paper.

11. Draw a freehand pictorial sketch of a typical PC mouse. Label your drawing and include approximate dimensions.

12. Sketch a block diagram to show how the components of a bicycle braking system are linked together. Label your diagram.

13. Sketch a flow chart to illustrate the process of fitting a mains plug to the lead of an electrical appliance. Include, in your flow chart, the correct choice and fitting of a mains fuse.

14. List THREE examples of schematic diagrams and explain where each is used.

15. With the aid of a sketch, explain how drawing zones are identified on a formal engineering drawing.

16. List FIVE items that usually appear in the title block of a formal engineering drawing.

17. Explain briefly what is meant by:

(a) A general arrangement (GA) drawing
(b) A detail drawing.

18. Sketch the conventional line style used in an engineering drawing to represent:

(a) The limit of a partial view
(b) Hidden edges and outlines
(c) Centre lines
(d) Extreme positions of movable parts.

19. Identify the projections shown in Figures 1.103 and 1.104.

20. Identify the line types marked A and B in Figure 1.105.

21. Identify each of the components shown in Figure 1.106.

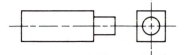

Figure 1.103 *See Question 19*

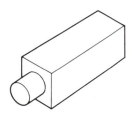

Figure 1.104 *See Question 19*

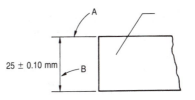

Figure 1.105 *See Question 20*

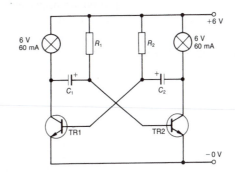

Figure 1.106 *See Question 21*

22. A flat square steel plate has sides measuring 220 mm. The plate has a round hole with a diameter of 110 mm cut in its exact centre. Use standard drawing conventions to show an orthographic view of this component. Label your drawing and include dimensions.

23. Refer to the drawing shown in Figure 1.107. What kind of drawing is this and what does the hatched area show?

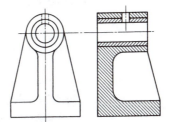

Figure 1.107 *See Question 23*

24. Refer to the drawing shown in Figure 1.108. Which one of the styles shown is correct for use in a formal engineering drawing?

(a)   Material: Steel
       Grind 0.5

(b)   MATERIAL: STEEL
       GRIND 0.5

(c)   *Material : Steel*
       Grind 0.5

(d)   MATERIAL: STEEL
       GRIND 0.5

Figure 1.108 *See Question 24*

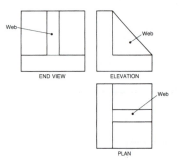

Figure 1.109 *See Question 26*

25. An engineering component comprises a metal alloy cube having sides measuring 100 mm. Draw this component using:

(a) Cavalier oblique projection
(b) Cabinet oblique projection
(c) Isometric projection.

26. Refer to the drawing shown in Figure 1.109. What type of projection has been used for the drawing and explain, briefly, how it was constructed.

27. Refer to the drawing shown in Figure 1.110. Redraw the component using third-angle projection.

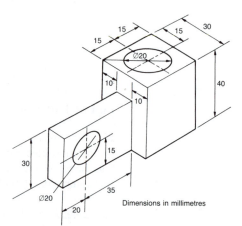

Figure 1.110 *See Question 27*

28. List the main section headings used in a technical report and briefly explain the purpose of each.

29. Sketch engineering drawing symbols that are used to indicate the following components:

(a) A 4/2 directional control valve
(b) A non-return valve
(c) A battery
(d) A variable resistor
(e) A semiconductor diode
(f) An iron-cored transformer.

30. Draw, using appropriate symbols, a two-way lighting circuit. Label your drawing clearly.

31. List FOUR advantages of using CAD in the preparation of engineering drawings compared with purely manual methods.

Chapter 2 Engineered products

Summary

This chapter covers Unit 2 of the GCSE engineering curriculum. It will show you how to make an engineered product. It will introduce you to some of the most commonly used engineering materials and components as well as the processes that are used in manufacturing. You will go on to apply this knowledge when you actually make an engineered product.

As you study this chapter you will learn the following:

- How to use a product specification
- How to read and interpret an engineering drawing
- How to select suitable materials, parts and components
- How to create a production plan
- How to use processes, tools and equipment to make an engineered product
- How to check that your work conforms to the required quality standard
- How to apply Health and Safety procedures.

As with the previous chapter, you should aim to build a *portfolio* of your work as you study this chapter. Later, your portfolio will provide you with a valuable collection of material that shows what you did and what you learned as you studied the chapter.

2.1 Introduction

Engineered products usually comprise assemblies of individual component parts. They can be divided into three main groups:

- *Mechanical products* such as gearboxes, pumps, engines and turbines.
- *Electronic products* such as mobile phones, audio systems and TV receivers.
- *Electromechanical devices* combine electrical and mechanical components and sub-assemblies and examples can range from washing machines to computer controlled machine tools.

We will start this chapter by taking a look at two engineered products: an electrician's screwdriver and a battery charger.

2.1.1 An electrician's screwdriver

Figure 2.1 shows an exploded view of a very simple engineered product, an electrician's screwdriver. This type of screwdriver combines the functions of a conventional screwdriver with the ability to indicate whether a mains circuit is live. It is worth considering what the design criteria might be for this particular product. The screwdriver must be strong enough to tighten and undo the small brass screws found in the terminals of electrical accessories. It must be insulated to withstand the potentials (voltages) met with in domestic, industrial and commercial installations. The current through the neon indicator lamp must be limited to a safe level under all conditions. It must be light in weight, compact and competitively priced.

These criteria can be met by careful selection of materials and manufacturing processes for the different *component parts* that go to make up the screwdriver:

- The blade should be made from a toughened medium carbon steel (0.8% carbon).
- The shank of the blade should be insulated using a polyvinylchloride (PVC) sleeve.
- The handle should be moulded from cellulose acetate (a tough, flame-resistant plastic with good insulating properties). The handle should be transparent so that the neon indicator lamp can be seen to light up.
- The blade should be moulded into the handle.
- The spring should be made from hard drawn phosphor bronze wire (a good electrical conductor that is also corrosion resistant).
- The end cap should be made from an aluminium alloy on a computer numerically controlled (CNC) lathe. This material is light in weight, easily cut, a good conductor and corrosion resistant. The process of manufacture is suitable for large batch production.
- The neon indicator lamp and the current limiting resistor should be *bought in* as standard components.

Key point
Material used for an engineered product must be selected for their 'fitness for purpose'.

All the materials chosen for the screwdriver should be readily available and relatively low in cost. They are selected for their *fitness for purpose*. Since screwdrivers are made and sold in large quantities, the manufacturing processes that we chose should be suitable for high volume batch production.

2.1.2 A power supply

Figure 2.2 shows the circuit for a power supply suitable for providing d.c. power to a variety of small electronic items. The output of the power supply is variable and can be adjusted over the range 3–35 V in two switched ranges. Figure 2.2 also shows some of the constructional details of the power supply. Most of the components for this product such as the transformer, rectifier pack, the sockets, switches,

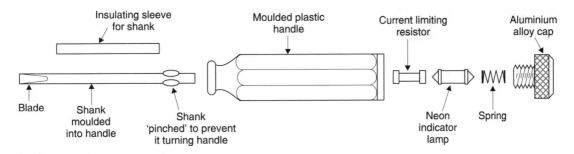

Figure 2.1 *An electrician's screwdriver*

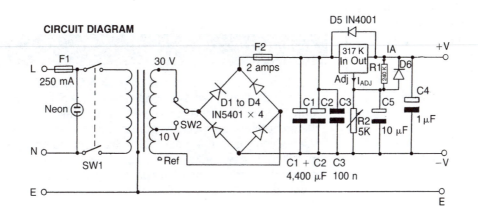

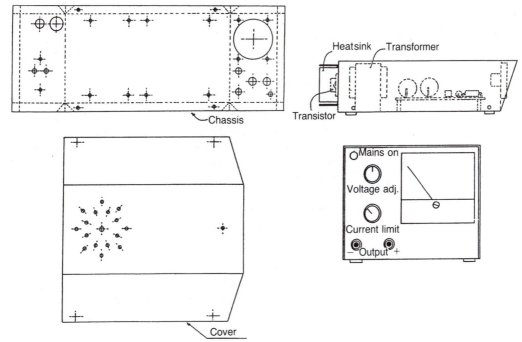

Figure 2.2 *A power supply*

Test your knowledge 2.1

Explain what a 'bought in' component is and why this type of component is sometimes preferred to one that is manufactured.

Test your knowledge 2.2

List THREE main factors that must be considered when choosing a material for a given engineering application.

fuse holders, etc., would be bought in. The power supply will have to have quite a high current output, so you must carefully consider how to keep the components cool. This particularly applies to the rectifier pack since solid-state devices are destroyed if they are allowed to overheat. For this reason an aluminium case should be used and a metal heat dissipator (or *heatsink*) should be fitted to the rectifier. In order to ensure that air flows inside the power supply's enclosure, cooling holes or louvres should be included in the design. These should be positioned so that moisture cannot enter the case. Since the equipment is connected to the mains supply and the case is of metal construction, the case must be earthed in accordance with current safety regulations.

Activity 2.1

Obtain a standard 13 A mains plug. Carefully take this apart and produce an exploded view of the plug clearly identifying each of the component parts. Make a full list of the component parts, naming each part and the material from which it is made (your tutor will be able to help you with this). Present your work on a single sheet of A4 or A3 paper using one of the blank drawing grids that you produced whilst studying Chapter 1.

2.2 Specifications

We have already briefly mentioned specifications in our chapter on Design and Graphical Communication. There we were concerned with the specifications that were used early in the design process to give a designer a very clear indication of the required performance of the product or service that's being designed.

In this chapter we are concerned with using a product specification to develop, and ultimately manufacture an engineered product or deliver an engineering service. At this point, it is important to remember that the information that goes to make up a full product specification often appears in both the written specification and also in the working drawings and diagrams (e.g. those that may specify dimensions or surface finish).

To ensure that the finished product or service meets the client's requirements, the specification for the product or service must be adequately detailed. The specification must also be something that is precise and measurable. For example, the specification for the electrician's screwdriver must, amongst other things, state the length of the shaft and the dimensions of the blade. In the case of the power supply, we would need to state the output voltage range (3–35 V in two switched ranges) and the maximum current that can be supplied (1 A). All of this is important information – not only to ensure that it is what the client wants and expects but also when the time comes to select components, materials and processes.

How to prepare a product specification

✓ Start by naming the product or service.

✓ Write a brief description of the product or service stating what it does as well as the design problem that it is intended to solve.

✓ List each of the main features of the product or service and say what the purpose or function of each feature is.

✓ Itemise the performance characteristics of the product or service and give a precise and measurable indication of each performance characteristic (such as indicated voltage range, maximum output current, etc.).

✓ List the physical characteristics of the product or service and give a precise and measurable indication of each characteristic (such as length, width, height, weight, etc.).

In order to make your product specification more useful you can

✓ Include relevant drawings and sketches.

✓ Include details of tolerances and the effect of external influences (such as temperature) on performance characteristics.

✓ Include features that relate to servicing and maintenance.

Activity 2.2

Take a look at the electrician's screwdriver that you met earlier in Section 2.1.1. Draw up a complete product specification for the screwdriver and present your work in the form of a single A4 page. Specify the materials for each part together with any relevant dimensions or finish (if in doubt, examine a real electrician's screwdriver). Add a title to your finished specification sheet and include it in your portfolio.

Activity 2.3

Take a look at the power supply that you met earlier in Section 2.1.2. Draw up a complete product specification for the power supply and present your work in the form of a single A4 page. Specify the materials or components used for each part and include a full electronic parts list. Also suggest dimensions and a finish for the enclosure (if in doubt, examine a real power supply). Add a title to your finished specification sheet and include it in your portfolio.

2.3
Production
planning

Having produced a detailed specification together with a comprehensive set of drawings we are ready to move on to the next stage, planning the production process. What we mean by this is the sequence of operations and processes that allow us to manufacture the product or, in the case of an engineered service, supply the service.

The sequence of operations is just as important as the processes that are involved. Also important is the choice of materials, components and parts. The production planning process involves considering a number of factors including:

- the available resources
- the materials, parts and components to be used
- the processes that are to be used
- the available equipment and machinery
- the sequence of production
- the arrangements for inspection and quality control
- Health and Safety factors.

To help you understand what production planning is about, let us assume that you have decided to construct a prototype battery-powered alarm that will emit a sound from a *piezoelectric transducer* whenever a bike is moved. The alarm is to be *set* (made active) and *reset* (deactivated) using a simple two-position *keyswitch*.

The movement of the bike is to be detected by means of a small *motion detector*. The motion detector is an electronic component that is mounted, together with the rest of the alarm circuit components, on a small *printed circuit board* (PCB).

The alarm circuit is to be powered from a *rechargeable battery* and the entire assembly is to be enclosed in a sealed (waterproof) *diecast aluminium alloy box* which will be secured to the bicycle frame by means of a bracket fitted with *anti-tamper bolts*.

The main components for the bike alarm are listed below:

- piezoelectric transducer
- keyswitch
- motion detector
- rechargeable battery
- battery holder
- printed circuit board*
- diecast enclosure
- clamp assembly.*

Note that you intend to manufacture the components marked * and to buy in all the other parts as they are available 'off-the-shelf'.

Since the PCB is going to provide a means of both mounting the electronic component and connecting them together, you will probably need to design and manufacture the PCB before doing anything else. Similarly, before you can assemble the PCB and the other components inside the case you will need to drill the diecast box so that the PCB, keyswitch and battery holder can be fitted to it. Deciding on the precise sequence of operations required to manufacture an engineered product during the manufacture of an engineered product can sometimes be quite a complex task. The task becomes more complicated when a large number of components or processes (or both) are involved. Imagine, for example, having to draw up a production schedule for a large aircraft! Fortunately, the production sequence for your bike alarm should be fairly straightforward and could typically be as follows:

1. Design and manufacture the PCB
2. Obtain electronic parts and solder them into PCB
3. Drill the diecast box to accommodate the PCB mounting pillars, keyswitch, piezoelectric transducer, battery holder, and clamping assembly
4. Assemble the PCB and other parts in the diecast box and solder any interconnecting wires that may be required
5. Charge the battery and insert into the battery holder
6. Test the alarm and check that it operates correctly
7. Manufacture the clamp assembly
8. Attach the clamp assembly to the diecast box.

Production planning is not just about the sequence of operations but, as we said earlier, it involves the selection of components and materials as well as the processes and equipment used in producing an engineered product or delivering an engineering service.

Let us think about some of the materials and components that you might use in the construction of your bike alarm. What would be the reason for using a diecast aluminium enclosure? What would be the reason for using a rechargeable battery? Why would you want to use anti-tamper bolts in the clamp assembly? What tools and equipment will you need to have available to manufacture the PC? All of these questions, and many more, need to be answered *before you start* to manufacture your product. This is the purpose of a production plan!

How to prepare a production plan

✓ Start by checking that you have a complete specification for your engineered product or service.

✓ Draw up a complete components and parts list and decide whether each component or part will be manufactured or bought-in.

✓ Decide on how you will manufacture any parts that are not going to be bought-in and what processes you will need to apply.

✓ Decide on what tools and equipment you have available (or will need to obtain) in order to manufacture the parts that are not to be bought-in.

✓ Decide on what materials and processes are most appropriate for manufacturing any parts that are not to be bought-in.

✓ Decide on what finish will be applied to the manufactured parts and components and how this will be applied.

✓ Decide on the most appropriate sequence of manufacture or assembly.

Test your knowledge 2.3

List FIVE factors that must be considered when preparing a production plan.

In order to make your production plan more useful you can

✓   Use sketches where appropriate to illustrate any processes that are non-standard.

✓   Use a flow chart to illustrate the sequence in which the processes are to be applied or the product is to be assembled.

✓   Include a timescale showing the time for each stage and an estimate of the total time to manufacture the product or to deliver the service.

✓   Include full details of the resources that you will require at each stage of production.

Activity 2.4

Prepare a production plan for the electrician's screwdriver that you met in Section 2.1.1. Think about the materials, parts and components that are to be used and whether any of these can be bought-in or need to be manufactured. Think about the tools and equipment that will be needed to manufacture the parts and also to assemble the product. Think also about how the screwdriver will be finished and how the finish will be applied. Identify the processes and activities that are needed during manufacture and suggest a sequence and timescale for each activity. Present your production plan as a series of notes and sketches covering each stage in the manufacture of the screwdriver. Add the completed production plan to your portfolio.

Activity 2.5

Prepare a production plan for the power supply that you met in Section 2.1.2. Think about the materials, parts and components that are to be used and whether any of these can be bought-in or need to be manufactured. Think about the tools and equipment that will be needed to manufacture the parts and also to assemble the product. Think also about how the power supply will be finished and how the finish will be applied. Identify the processes and activities that are needed during manufacture and suggest a sequence and timescale for each activity. Present your production plan as a series of notes and sketches covering each stage in the manufacture of the power supply. Add the completed production plan to your portfolio.

2.3.1 *Quality control and quality assurance*

Quality is usually defined as *fitness for purpose*, a measure of the degree to which a product conforms to specification and standards of workmanship. Quality is a key objective for most engineering companies.

As quality can often be somewhat subjective (and is often more about the customer's perception rather than accurate measurements of performance) it is important to use a set of more objective criteria for assessing 'fitness for purpose' which usually include:

- design quality
- conformance quality
- reliability
- maintainability.

Design quality
Design quality relates to the development of a specification for the product that meets a customer's identified needs. Design quality is usually the joint responsibility of a company's marketing or customer liaison function and its research and development function.

Conformance quality
Conformance quality is about producing a product that conforms to the design specification. A product that conforms is a quality product, even if the design itself is for a product that might be considered 'cheap'. If this seems a little contradictory, consider the following example.

A specification is drawn up for a disposable 35 mm camera. As this device will only be used once this can be made from inexpensive materials and can use a low-cost plastic lens. The camera has a built-in flash unit but this since this only needs to be used for a maximum of 15 exposures, the battery does not need to have a high capacity. As long as the budget camera conforms to the design specification (agreed with the customer) and, even though the camera is inferior when compared with most other cameras, it can still be considered to be a high-quality product.

Reliability
Reliability is often measured in terms of the *mean time to failure (MTTF)*. On average, a product (e.g. an electric light bulb) will operate for a specified time (its MTBF) before it fails. If the product is more complex and repairable (e.g. a TV set) we usually measure reliability in terms of *mean time between failures (MTBF)*. The greater the value of MTTF or MTBF the greater the reliability of a product however it is important to remember that the figures quoted are *average* values and there is nothing to prevent a light bulb or a TV set failing well in advance (or well after) its MTTF/MTBF.

Maintainability
Maintainability is a term that is often used to describe the ease of which a product can be repaired if and when it does becomes faulty. Maintainability may also require that routine maintenance is performed (e.g. most modern cars require that a service is carried out after 9000 to 12,000 miles or within 1 year, whichever occurs first).

Without routine maintenance (where lubricants and other components that may wear out can be replaced) reliability is likely to be very significantly reduced.

Quality assurance

Quality assurance is concerned with *all four* of the aspects listed earlier. The activities that make up a quality assurance system include the following:

- inspection, testing and checking of incoming materials and components
- inspection, testing and checking of the company's own products
- administering any supplier quality assurance systems
- dealing with complaints and warranty failures
- building quality into the manufacturing process.

Whilst many of these activities are performed in order to monitor quality *after* the event, others may be carried out to prevent problems *before* they occur and some may be carried out to determine the causes of failure that relate to design rather than manufacturing faults.

2.3.2 Health and safety

When planning the production process it is important to take into account specific factors relating to Health and Safety. Many engineering processes are potentially hazardous and these include activities such as casting, cutting, soldering, welding, etc. In addition, some processes may involve the use of hazardous materials and chemicals. Even the most basic and straightforward activities can potentially be dangerous if carried out using inappropriate tools, materials and methods.

In all cases, the correct tools and protective equipment should be used and proper training should be provided. In addition, safety warnings and notices should be prominently placed in the workplace and access to areas where hazardous processes take place should be restricted and carefully controlled so that only appropriately trained personnel can be present. In addition, the storage of hazardous materials (chemicals, radioactive substances, etc.) requires special consideration and effective access control.

2.4 Choosing materials, parts and components

By now, you should be beginning to appreciate that a very wide range of materials, parts and components is used in engineering! Figure 2.3 shows the main groups of engineering materials. The Latin name for iron is *ferrum*, so it is not surprising that ferrous metals and alloys are all based on the metal iron. Alloys consist of two or more metals (or metals and non-metals) that have been brought together as compounds or solid solutions to produce a metallic material with special properties. For example, an alloy of iron, carbon, nickel and chromium is stainless steel. This is a corrosion-resistant ferrous alloy. Non-ferrous metals and alloys are the rest of the metallic materials available. Non-metals can be natural, such as rubber, or they can be synthetic such as the plastic compound PVC.

Key point

Alloys are compounds or solid solutions of two or more metals (or metals and non-metals).

Test your knowledge 2.4

What is stainless steel and which metals are used in its production?

Test your knowledge 2.5

Name ONE natural non-metal and ONE synthetic non-metal used in engineering.

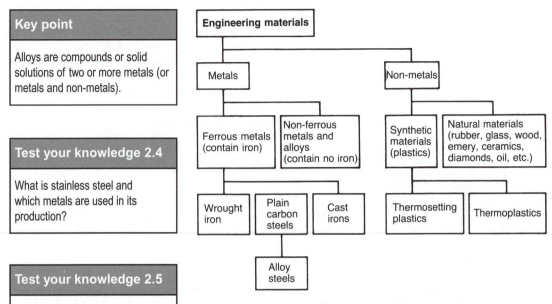

Figure 2.3 *The main groups of engineering materials*

When selecting a material you need to make certain that it has properties that are appropriate for the job it has to do. For example, we must ask ourselves the following questions:

- Will it corrode in its working environment?
- Will it weaken or melt in a hot environment?
- Will it break under normal working conditions?
- Can it be easily cast, formed or cut to shape?

You can assess the suitability of different materials for a particular engineering application by comparing their properties. The properties of a material used in engineering are extremely important and careful consideration is usually given to this aspect of production planning.

2.4.1 Chemical properties

When engineers look at the chemical properties of a material they are usually concerned with two things: corrosion and degradation.

Corrosion
This is caused by the metals and metal alloys being attacked and eaten away by chemical substances. For example, the rusting of ferrous metals and alloys is caused by the action of atmospheric oxygen in the presence of water. Another example is the attack on aluminium and some of its alloys by strong alkali solutions. Take care when using degreasing agents on such metals. Copper and copper based alloys are stained and corroded by the active sulphur and chlorine products found in some heavy duty cutting lubricants.

Key point

Engineers are usually concerned with corrosion and degradation when they look at the chemical properties of a material.

Degradation

Non-metallic materials do not corrode but they can be attacked by chemical substances. Since this weakens or even destroys the material it is referred to as degradation. Unless specially compounded, rubber is attacked by prolonged exposure to oil. Synthetic (plastic) materials can be softened by chemical solvents. Exposure to the ultraviolet rays of sunlight can weaken (perish) rubbers and plastics unless they contain compounds that filter out such rays.

2.4.2 Electrical properties

The electrical properties of a material are important when selecting it for an engineering application that requires the material to either allow or prevent the flow of an electric current to some degree. We specify this property by referring to its electrical resistance.

> **Key point**
>
> Materials that have a low electrical resistance are described as 'good conductors'.

> **Key point**
>
> Materials that have a very high electrical resistance are described as 'insulators'.

Electrical resistance

Materials with a very low resistance to the flow of an electric current are good electrical *conductors*. Materials with a very high resistance to the flow of electric current are good *insulators*. Generally, metals are good conductors and non-metals are good insulators (poor conductors). A notable exception is carbon which conducts electricity despite being a non-metal. The electrical resistance of a metal conductor depends upon:

- Its length (the longer it is the greater its resistance)
- Its thickness (the thicker it is the lower its resistance)
- Its temperature (the higher its temperature the greater is its resistance)
- Its resistivity (this is the resistance measured between the opposite faces of a metre cube of the material).

Note that a small number of non-metallic materials, such as silicon, have atomic structures that fall between those of electrical conductors and insulators. These materials are called semi-conductors and are used for making solid-state devices such as transistors.

2.4.3 Magnetic properties

All materials respond to strong magnetic fields to some extent. Only the ferromagnetic materials respond sufficiently to be of interest. The more important ferromagnetic materials are the metals iron, nickel and cobalt. Magnetic materials are often divided into two types: soft and hard.

Soft magnetic materials, such as soft iron, can be magnetised by placing them in a magnetic field. They cease to be magnetised as soon as the field is removed.

Hard magnetic materials, such as high-carbon steel that has been hardened by cooling it rapidly (quenching) from red heat, also become magnetised when placed in a magnetic field. Hard magnetic materials retain their magnetism when the field is removed. They become permanent magnets.

Permanent magnets can be made more powerful for a given size by adding cobalt to the steel to make an alloy. Soft magnetic materials can be made more efficient by adding silicon or nickel to the pure iron. Silicon-iron alloys are used for the rotor and stator cores of electric motors and generators. Silicon-iron alloys are also used for the cores of power transformers.

2.4.4 Thermal properties

Thermal properties are to do with how a material responds to heat and different temperatures. Thermal properties include:

Melting temperature
The melting temperature of a material is the temperature at which a material loses its solid properties. Most plastic materials and all metals become soft and eventually melt. Note that some plastics do not soften when heated, they only become charred and are destroyed. This will be considered later in this chapter.

Thermal conductivity
This is the ease with which materials conduct heat. Metals are good conductors of heat. Non-metals are poor conductors of heat. Therefore, non-metals are heat insulators.

Expansion
Metals expand appreciably when heated and contract again when cooled. They have high coefficients of linear expansion. Non-metals expand to a lesser extent when heated. They have low coefficients of linear expansion. Again, these properties will be considered in more detail in your science unit.

2.4.5 Mechanical properties

Mechanical properties are important when engineering materials are to withstand loads and forces imposed on them.

Strength
This is the ability of a material to resist an applied force (load) without fracturing (breaking). It is also the ability of a material not to *yield*. Yielding is when the material 'gives' suddenly under load and changes shape permanently but does not break. This is what happens when metal is bent or folded to shape. The load or force can be applied in various ways as shown in Figure 2.4.

You must be careful when interpreting the strength data quoted for various materials. A material may appear to be strong when subjected to a static load, but will break when subjected to an impact load. Materials also show different strength characteristics when the load is applied quickly than when the load is applied slowly.

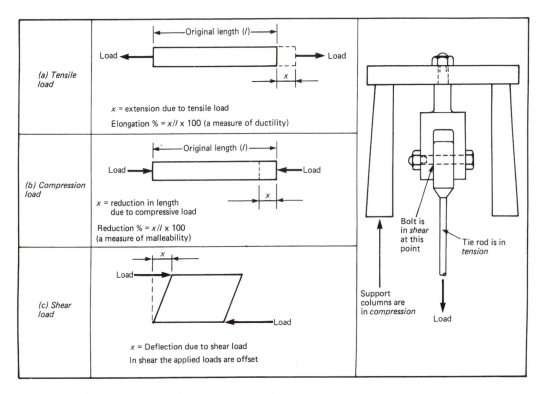

Figure 2.4 *Different ways in which a load can be applied*

Toughness

This is the ability of a material to resist impact loads as shown in Figure 2.5. Here, the toughness of a piece of high-carbon steel in the soft (annealed) condition is compared with a piece of the same steel after it has been hardened by raising it to red-heat and cooling it quickly (*quenching* it in cold water). The hardened steel shows a greater strength, but it lacks toughness.

A test for toughness, called the Izod test, uses a notched specimen that is hit by a heavy pendulum (see Figure 2.5). The test conditions are carefully controlled, and the energy absorbed in bending or breaking the specimen is a measure of the toughness of the material from which it was made.

Elasticity

Materials that change shape when subjected to an applied force but spring back to their original size and shape when that force is removed are said to be elastic. They have the property of elasticity.

Plasticity

Materials that flow to a new shape when subjected to an applied force and keep that shape when the applied force is removed are said to be plastic. They have the property of plasticity.

Ductility

Materials that can change shape by plastic flow when they are subjected to a pulling (tensile) force are said to be ductile. They have the property of ductility. This is shown in Figure 2.6(a).

Test your knowledge 2.6

Explain what happens when a material 'yields' when a load applies to it.

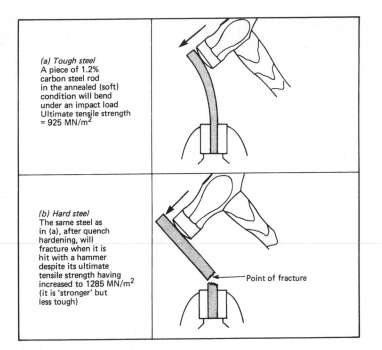

(a) Tough steel
A piece of 1.2%
carbon steel rod
in the annealed (soft)
condition will bend
under an impact load
Ultimate tensile strength
= 925 MN/m^2

(b) Hard steel
The same steel as
in (a), after quench
hardening, will
fracture when it is
hit with a hammer
despite its ultimate
tensile strength having
increased to 1285 MN/m^2
(it is 'stronger' but
less tough)

Point of fracture

Figure 2.5 *Impact loads*

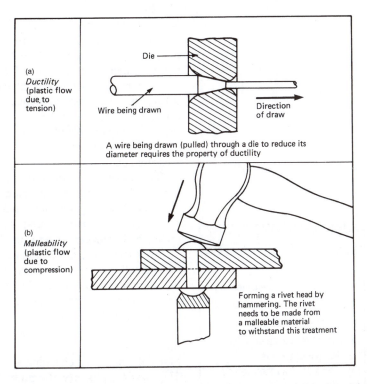

(a)
Ductility
(plastic flow
due to
tension)

Die

Wire being drawn

Direction
of draw

A wire being drawn (pulled) through a die to reduce its
diameter requires the property of ductility

(b)
Malleability
(plastic flow
due to
compression)

Forming a rivet head by
hammering. The rivet
needs to be made from
a malleable material
to withstand this treatment

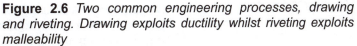

Figure 2.6 *Two common engineering processes, drawing
and riveting. Drawing exploits ductility whilst riveting exploits
malleability*

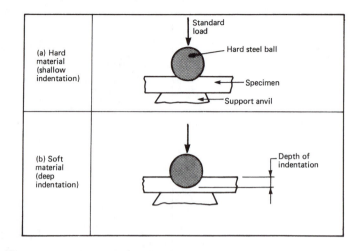

Figure 2.7 *The effect of pressing a hard steel ball into two materials with different hardness problems*

Malleability
Materials that can change shape by plastic flow when they are subjected to a squeezing (compressive) force are said to be malleable. They have the property of malleability. This is shown in Figure 2.6(b).

Hardness
Materials that can withstand scratching or indentation by an even harder object are said to be hard. They have the property of hardness. Figure 2.7 shows the effect of pressing a hard steel ball into two pieces of metal with the same force. The ball sinks further in to the softer of the two pieces of metal than it does into the harder. There are various hardness tests available. The Brinell hardness test uses the principles set out above. A hardened steel ball is pressed into the specimen by a controlled load. The diameter of the indentation is measured using a special microscope. The hardness number is obtained from the measured diameter by use of conversion tables.

The Vickers test is similar but uses a diamond pyramid instead of a hard steel ball. This enables harder materials to be tested. The diamond pyramid leaves a square indentation and the diagonal distance across the square is measured. Again, conversion tables are used to obtain the hardness number from the measured distance.

The Rockwell test uses a diamond cone. A minor load is applied and a small indentation is made. A major load is then added and the indentation increases in depth. This increase in depth of the indentation is directly converted into the hardness number and it can be read from a dial on the machine.

Rigidity
Materials that resist changing shape under load are said to be rigid. They have the property of rigidity. The opposite of rigidity is flexibility. Rigid materials are usually less strong than flexible materials. For example, cast iron is more rigid than steel but steel is the stronger and tougher. However, the rigidity of cast iron makes it a useful material

for machine frames and beds. If such components were made from a more flexible material the machine would lack accuracy and it would be deflected by the cutting forces.

2.4.6 Metals

Metals are widely used in the manufacture of engineered products. Metals can be divided into two main types: ferrous and non-ferrous. Ferrous metals include iron and steel whilst non-ferrous metals include copper, brass and aluminium.

As previously stated, ferrous metals are based upon the metal iron. For engineering purposes iron is usually associated with various amounts of the non-metal carbon. When the amount of carbon present is less than 1.8% we call the material steel. The figure of 1.8% is the theoretical maximum. In practice, there is no advantage in increasing the amount of carbon present above 1.4%. We are only going to consider the plain carbon steels. Alloy steels are beyond the scope of this book. The effects of the carbon content on the properties of plain carbon steels are shown in Figure 2.8.

Cast irons are also ferrous metals. They have substantially more carbon than the plain carbon steels. Grey cast irons usually have a carbon content between 3.2% and 3.5%. Not all this carbon can be taken up by the iron and some is left over as flakes of graphite

> **Key point**
>
> Ferrous metals are based on iron.

> **Key point**
>
> Steel is iron with less than 1.8% carbon present.

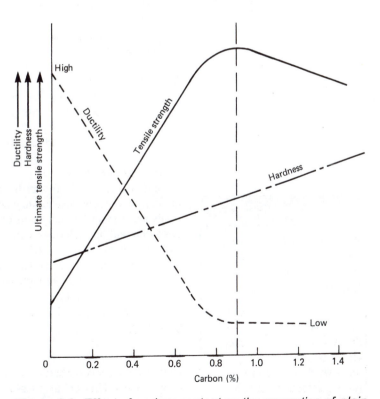

Figure 2.8 *Effect of carbon content on the properties of plain carbon steels*

Name	Group	Carbon content (%)	Some uses
Dead mild steel (low carbon steel)	Plain carbon steel	0.10–0.15	Sheet for pressing out components such as motor car body panels. General sheet-metal work. Thin wire, rod and drawn tubes
Mild steel (low carbon steel)	Plain carbon steel	0.15–0.30	General purpose workshop rod, bars and sections. Boiler plate. Rolled steel beams, joists, angles, etc.
Medium carbon steel	Plain carbon steel	0.30–0.50	Crankshafts, forgings, axles, and other stressed components
		0.50–0.60	Leaf springs, hammer heads, cold chisels, etc.
High carbon steel	Plain carbon steel	0.8–1.0 1.0–1.2 1.2–1.4	Coil springs, wood chisels Files, drills, taps and dies Fine-edge tools (knives, etc.)
Grey cast iron	Cast iron	3.2–3.5	Machine castings

Table 2.1 *Ferrous metals*

between the crystals of metal. It is these flakes of graphite that gives cast iron its particular properties and makes it a 'dirty' metal to machine. The compositions and typical uses of plain carbon steels and a grey cast iron are summarised in Table 2.1.

Low carbon steels
Low carbon steels (also called mild steels) are the cheapest and most widely used group of steels. Although they are the weakest of the steels, nevertheless they are stronger than most of the non-ferrous metals and alloys. They can be hot and cold worked and machined with ease.

Medium carbon steels
Medium carbon steels are harder, tougher, stronger and more costly than the low carbon steels. They are less ductile than the low carbon steels and cannot be bent or formed to any great extent in the cold condition without risk of cracking. Greater force is required to bend and form them. Medium carbon steels hot forge well but close temperature control is essential. Two carbon ranges are shown. The lower carbon range can only be toughened by heating and quenching (cooling quickly by dipping in water). They cannot be hardened. The higher-carbon range can be hardened and tempered by heating and quenching.

High-carbon steels
High-carbon steels are harder, stronger and more costly than medium-carbon steels. They are also less tough. High-carbon steels are available as hot-rolled bars and forgings. Cold drawn high-carbon steel wire (piano wire) is available in a limited range of sizes. Centreless

Test your knowledge 2.10

Name TWO tests of a material's hardness.

ground high-carbon steel rods (silver steel) are available in a wide range of diameters (inch and metric sizes) in lengths of 333 mm, 1 m, and 2 m. High-carbon steels can only be bent cold to a limited extent before cracking. They are mostly used for making cutting tools such as files, knives and carpenters' tools.

2.4.7 Non-ferrous metals and alloys

Non-ferrous metals (i.e. metals that are *not* based on iron) include metals such as aluminium and zinc as well as alloys such as brass and bronze. We shall start by looking at copper-a material that is widely used in engineering.

High-conductivity copper
High-conductivity copper is better than 99.9% pure and it is widely used for electrical conductors and switchgear components. It is second only to silver in conductivity but it is much more plentiful and very much less costly. Pure copper is too soft and ductile for most mechanical applications.

Tough pitch copper
Tough pitch copper is suitable for general purpose applications such as roofing, chemical plant, decorative metal work, and copper-smithing, tough pitch copper is used. This contains some copper oxide which makes it stronger, more rigid and less likely to tear when being machined. As it is not so highly refined, it is less expensive than high conductivity copper.

There are many other grades of copper for special applications. Copper is also the basis of many important alloys such as brass and bronze and we will be considering these next. The general properties of copper are:

- Relatively high strength.
- Very ductile so that it is usually cold worked. An annealed (softened) copper wire can be stretched to nearly twice its length before it snaps.
- Corrosion resistant.
- Second only to silver as a conductor of heat and electricity.
- Easily joined by soldering and brazing. For welding, a phosphorous deoxidised grade of copper must be used.

Copper is a available as cold-drawn rods, wires and tubes. It is also available as cold-rolled sheet, strip and plate. Hot worked copper is available as extruded sections and hot stampings. It can also be cast. Copper powders are used for making sintered components. It is one of the few pure metals of use to the engineer as a structural material.

Brass
Brass, an *alloy* of copper and zinc, is another non-ferrous metal. The properties of a brass alloy and the applications for which you can use it depends upon the amount of zinc present. Most brasses are attacked

by seawater. The salt water eats away the zinc (dezincification) and leaves a weak, porous, spongy mass of copper. To prevent this happening, a small amount of tin is added to the alloy. There are two types of brass that can be used at sea or on land near the sea. These are Naval Brass and Admiralty Brass.

Brass is a difficult metal to cast and brass castings tend to be coarse grained and porous. Brass depends upon hot rolling from cast ingots, followed by cold rolling or drawing to give it its mechanical strength. It can also be hot extruded and plumbing fittings are made by hot stamping. Brass machines to a better finish than copper as it is more rigid and less ductile than that metal. Table 2.2 lists some typical brasses, together with their compositions, properties and applications.

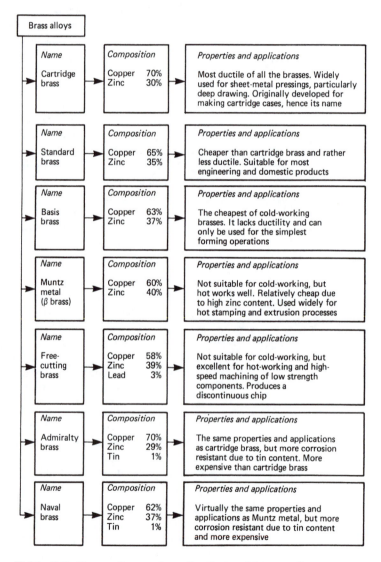

Table 2.2 *Properties and applications of brass alloys*

Tin bronzes

As the name implies, the tin bronzes are alloys of copper and tin. These alloys also have to have a deoxidising element present to prevent the tin from oxidising during casting and hot working. If the tin oxidises the metal becomes hard and 'scratchy' and is weakened.

The two deoxidising elements commonly used are:

- zinc in the gun-metal alloys
- phosphorus in the phosphor bronze alloys.

Unlike the brass alloys, the bronze alloys are usually used as castings. However, low-tin content phosphor bronze alloys can be extensively cold worked. Tin bronze alloys are extremely resistant to corrosion and wear and are used for high-pressure valve bodies and heavy-duty bearings. Table 2.3 lists some typical bronze alloys together with their compositions, properties and applications.

Test your knowledge 2.11

An alloy consists of 94% copper with 0.2% phosphorus and 5.8% tin. Name this alloy.

Test your knowledge 2.12

State the composition of each of the following alloys:

(a) Standard brass.
(b) Naval brass.
(c) Admiralty gun metal.

Test your knowledge 2.13

Select a suitable non-ferrous metal for each of the following, giving reasons for your choice.

(a) The body casting of a water pump.
(b) Screws for clamping the electric cables in the terminals of a domestic electric light switch.
(c) A bearing bush.
(d) A deep drawn, cup-shaped component for use on land.
(e) A ship's fitting made by hot stamping.

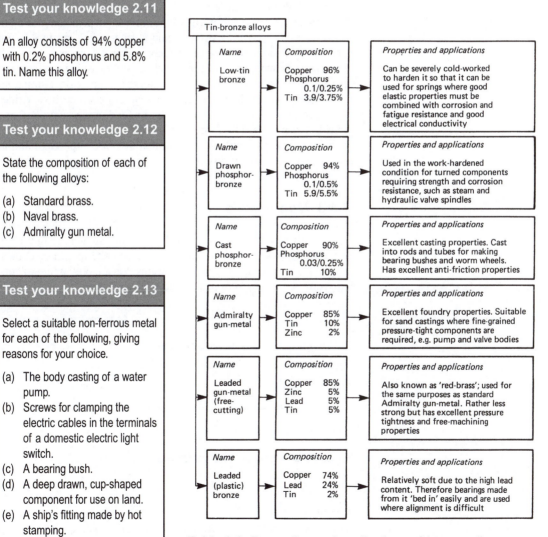

Table 2.3 *Properties and applications of bronze alloys*

Aluminium

Aluminium has a density approximately one-third that of steel. However, it is also very much weaker so its strength/weight ratio is inferior. For stressed components, such as those found in aircraft, aluminium alloys have to be used. These can be as strong as steel and nearly as light as pure aluminium.

High purity aluminium

This is second only to copper as a conductor of heat and electricity. It is very difficult to join by welding or soldering and aluminium conductors are often terminated by crimping. Despite these difficulties, it is increasingly used for electrical conductors where its light weight and low cost compared with copper is an advantage. Pure aluminium is resistant to normal atmospheric corrosion but it is unsuitable for marine environments. It is available as wire, rod, cold-rolled sheet and extruded sections for heatsinks.

Commercially pure aluminium

This is not as pure as high-purity aluminium and it also contains up to 1% silicon to improve its strength and stiffness. As a result it is not such a good conductor of electricity nor is it so corrosion resistant. It is available as wire, rod, cold-rolled sheet and extruded sections and also as castings and forgings. Being stiffer than high-purity aluminium it machines better with less tendency to tear. It forms non-toxic oxides on its surface which makes it suitable for food processing plant and utensils. It is also used for forged and die-cast small machine parts. Due to their range and complexity, the light alloys based upon aluminium are beyond the scope of this course.

2.4.8 Non-metals

Non-metals can be grouped under the headings shown in Figure 2.9. In addition, wood is also used for making the patterns which, in turn, are used in producing moulds for castings. We are only going to consider some ceramics, thermosets and thermoplastics.

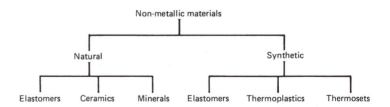

Figure 2.9 *Non-metallic materials*

Ceramics

The word ceramic comes from a Greek work meaning potter's clay. Originally, ceramics referred to objects made from potter's clay.

Nowadays, ceramic technology has developed a range of materials far beyond the traditional concepts of the potter's art. These include:

- Glass products
- Abrasive and cutting tool materials
- Construction industry materials
- Electrical insulators
- Cements and plasters for investment moulding
- Refractory (heat resistant) lining for furnaces
- Refractory coatings for metals.

The four main groups of ceramics and some typical applications are summarised in Table 2.4.

Test your knowledge 2.14

List THREE applications for ceramic materials.

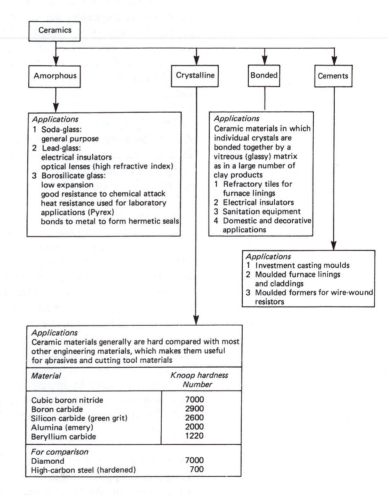

Table 2.4 *Applications of ceramic materials*

The common properties of ceramic materials can be summarised as follows:

Strength
Ceramic materials are reasonably strong in compression, but tend to be weak in tension and shear. They are brittle and lack ductility.

They also suffer from micro-cracks that occur during the firing process. These lead to fatigue failure. Many ceramics retain their high compressive strength at very high temperatures.

Hardness
Most ceramic materials are harder than other engineering materials as shown in Table 2.4. They are widely used for cutting tool tips and abrasives. They retain their hardness at very high temperatures that would destroy high-carbon and high-speed steels. However, they have to be handled carefully because of their brittleness.

Refractoriness
This is the ability of a material to withstand high temperatures without softening and deforming under normal service conditions. Some refractories such as high-alumina brick and fireclays tend to soften gradually and may collapse at temperatures well below their fusion (melting) temperatures. Refractories made from clays containing a high proportion of silica to alumina are most widely used for furnace linings.

Electrical properties
As well as being used for weather-resistant high-voltage insulators for overhead cables and sub-station equipment, ceramics are now being used for low-loss high-frequency insulators. For example, they are being used for the dielectric in silvered ceramic capacitors for high-frequency applications.

In all the above examples the ceramic material is poly-crystalline. That is the material is made up of a lot of very tiny crystals. For solid-state electronic devices single crystals of silicon are grown under very carefully controlled conditions. The single crystal can range from 50 to 150 mm diameter with a length ranging from 500 to 2500 mm. These crystals are without impurities. They are then cut up into thin wafers and made into such devices as thermistors, diodes, transistors and integrated circuits.

This is done by doping the pure silicon wafers with small, controlled amounts of carefully selected impurities. Some impurities give the silicon n-type characteristics. That is they make the silicon electrically negative by increasing the number of electrons present. Some impurities give the silicon p-type characteristics. That is they make the silicon electrically positive by reducing the number of electrons present.

Thermosetting plastics
Themosetting plastics are also known as *thermosets*. These materials are available in powder or granular form and consist of a synthetic resin mixed with a *filler*. The filler reduces the cost and modifies the properties of the material. A colouring agent and a lubricant is also added. The lubricant helps the plasticised moulding material to flow into the fine detail of the mould. The moulding material is subjected to heat and pressure in the moulds during the moulding process. The hot moulds not only plasticise the moulding material so that it flows into all the detail of the moulds, the heat also causes a chemical change in the material. This chemical change is called polymerisation

or, more simply, *curing*. Once cured, the moulding is hard and rigid. It can never again be softened by heating. If made hot enough it will just burn. Some thermosets and typical applications are summarised in Table 2.5.

Type	Applications
Phenolic resins and powders	The original 'Bakelite' type of plastic materials, hard, strong and rigid. Moulded easily and heat 'cured' in the mould. Unfortunately, they darken during processing and are only available in the darker and stronger colours. Phenolic resins are used as the 'adhesive' in making plywoods and laminated plastic materials (Tufnol)
Amino (containing nitrogen) resins and powders	The basic resin is colourless and can be coloured as required. Can be strengthened by paper-pulp fillers and are suitable for thin sections. Used widely in domestic electrical switchgear
Polyester resins	Polyester chains can be cross-linked by adding monomer such as styrene, when the polyester ceases to behave as a thermoplastic and becomes a thermoset. Curing takes place by internal heating due to chemical reaction and not by heating the mould. Used largely as the bond in the production of glass fibre mouldings
Epoxy resins	The strongest of the plastic materials used widely as adhesives, can be 'cold cast' to form electrical insulators and used also for potting and encapsulating electrical components

Table 2.5 *Thermosetting plastics*

Thermoplastics

Unlike the thermosets we have just considered, thermoplastics soften every time they are heated. In fact, any material trimmed from the mouldings can be ground up and recycled. They tend to be less rigid but tougher and more 'rubbery' than the thermosetting materials. Some thermoplastics and typical applications are summarised in Table 2.6.

Reinforced plastics

The strength of plastics can be increased by reinforcing them with fibrous materials. There are two main types: laminated plastics and glass reinforced plastics (GRP).

Laminated plastics (Tufnol)

Fibrous material, such as paper, woven cloth, woven glass fibre, etc., is impregnated with a thermosetting resin. The sheets of impregnated material are laid up in powerful hydraulic presses and they are heated and squeezed until they become solid and rigid sheets, rods, tubes, etc. This material has a high strength and good electrical properties. It can be machined with ordinary metal working tools and machines. Tufnol is used for making insulators, gears and bearing bushes.

Glass reinforced plastics

Woven glass fibre and chopped strand mat can be bonded together by polyester or by epoxy resins to form mouldings. These may range

Type	Material	Characteristics
Acrylics	Polymethyl-methacrylate	Materials of the 'Perspex' or 'Plexiglass' types. Excellent light transmission and optical properties, tough, non-splintering and can be easily heat-bent and shaped. Excellent high-frequency electrical insulators
Cellulose plastics	Nitro-cellulose	Materials of the 'celluloid' type. Tough, waterproof, and available as preformed sections, sheets and films. Difficult to mould because of their high flammability. In powder form nitro-cellulose is explosive
	Cellulose acetate	Far less flammable than nitro-cellulose and the basis of photographic 'safety' film. Frequently used for moulded handles for tools and electrical insulators
Fluorine plastics (Teflon)	Polytetrafluoro-ethylene (PTFE)	A very expensive plastic material, more heat resistant than any other plastic. Also has the lowest coefficient of friction. PTFE is used for heat-resistant and anti-friction coatings. Can be moulded (with difficulty) to produce components with a waxy feel and appearance
Nylon	Polyamide	Used as a fibre or as a wax-like moulding material. Tough, with a low coefficient of friction. Cheaper than PTFE but loses its strength rapidly when raised above ambient temperature. Absorbs moisture readily, making it dimensionally unstable and a poor electrical insulator
Polyesters (Terylene)	Polyethylene-teraphthalate	Available as a film or in fibre form. Ropes made from polyesters are light and strong and have more 'give' than nylon ropes. The film makes an excellent electrical insulator
Vinyl plastics	Polythene	A simple material, relatively weak but easy to mould, and a good electrical insulator. Used also as a waterproof membrane in the building industry
	Polypropylene	A more complicated material than polythene. Can be moulded easily and is similar to nylon in many respects. Its strength lies between polythene and nylon. Cheaper than nylon and does not absorb water
	Polystyrene	Cheap and can be easily moulded. Good strength but tends to be rigid and brittle. Good electrical insulation properties but tends to craze and yellow with age
	PVC	Tough, rubbery, practically non-flammable, cheap and easily manipulated. Good electrical properties and used widely as an insulator for flexible and semi-flexible cables

Table 2.6 *Thermoplastic materials*

from simple objects such as crash helmets to complex hulls for ocean going racing yachts. The thermosetting plastics used are cured by chemical action at room temperature and a press is not required. The glass fibre is laid up over plaster or wooden moulds and coated with the resin which is well worked into the reinforcing material. Several layers or *plys* may be built up according to the strength required. When cured the moulding is removed from the mould. The mould can be used again. Note that the mould is coated with a *release agent* before moulding commences.

Although the properties of plastic materials can vary widely, they all have some general properties in common.

Strength/weight ratio

Plastic materials vary considerably in strength and some of the stronger (such as nylon) compare favourably with the metals. All plastics have a lower density than metals and, therefore, chosen with care and proportioned correctly to their strength/weight ratio compares favourably with the light alloys.

Corrosion resistance

Plastic materials are inert to most inorganic chemicals and some are inert to all solvents. Thus they can be used in environments that are hostile to the most corrosion-resistant metals and many naturally occurring non-metals.

Electrical resistance

All plastic materials are good electrical insulators, particularly at high frequencies. However, when compared with ceramic materials, their usefulness is limited by their softness and low heat resistance. Flexible plastics such as PVC are useful for the insulation and protection of flexible electric cables.

Activity 2.6

Examine and dismantle a typical garden lawn sprinkler of the rotary type. Sketch the components for identification purposes (detail not required) and choose a suitable material for each component giving reasons for your choice. Present your work in the form of a one-page 'fact sheet'.

Activity 2.7

An electrical equipment manufacturer has asked you to advise them on the selection of materials to be used in the manufacture of an adjustable desk lamp which is to be fitted with a switch and is to accept a conventional 40 W light bulb. Sketch a suitable design and list each of the component parts required. Suggest, with reasons, a material to be used for each component. Present your work in the form of an illustrated word processed report.

Wherever possible, rather than manufacture from raw materials, you should use standard commercially available components. As these are mass produced in very large quantities, their cost is kept to a minimum. If they are made to an internationally acceptable standard, their quality is guaranteed. Let us now take a look at some of the more widely used mechanical and electrical components.

2.4.9 Mechanical components

Mechanical components include such common items as screwed fastenings and riveted joints. Since these components are very widely used, it is worth spending a little time explaining what they do and how they work:

Screwed fastenings
Screwed fastenings refer to nuts, bolts, screws and studs. These come in a wide variety of sizes and types of screw thread. When selecting a screwed fastening for any particular purpose, you should ask yourself the following questions:

- Is the fastening strong enough for the application?
- Is the material from which the fastening is made corrosion resistant under service conditions and is it compatible with the metals being joined?
- Is the screw thread chosen, suitable for the job? Coarse threads are stronger than fine threads, particularly in soft metals, such as aluminium. Fine threads are less likely to work loose.

Figure 2.10 shows some typical screwed fastenings and it also shows how they are used.

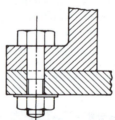

(a) Section through a bolted joint (plain shank extends beyond joint face)

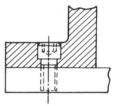

(b) Cap head socket screw (head recessed into counterbore to provide flush surface)

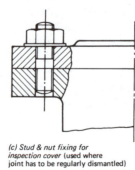

(c) Stud & nut fixing for inspection cover (used where joint has to be regularly dismantled)

(d) Cheese head brass screw (for clamping electrical conductor into terminal)

Figure 2.10 *Some typical screwed fastenings*

There are a large variety of heads for screwed fastenings, and the selection is usually a compromise between strength, appearance and ease of tightening. The hexagon head is usually selected for general engineering applications. The more expensive cap-head screw is widely used in the manufacture of machine tools, jigs and fixtures, and other highly stressed applications. These fastenings are forged from high-tensile alloy steels, thread rolled and heat treated. By recessing the cap-head, a flush surface is provided for safety and easy cleaning. Figure 2.11 shows some alternative screw heads.

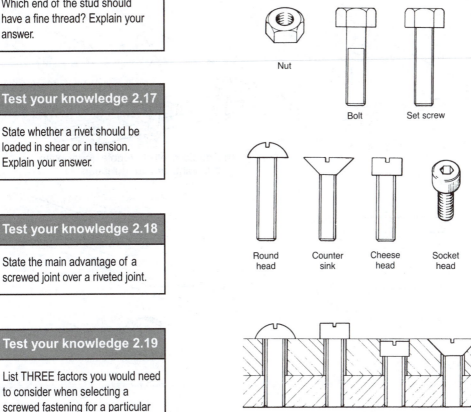

Applications of screw heads

Figure 2.11 *Various types of nut, bolt and screw*

Riveted joints

Figure 2.12 shows some typical riveted joints. Riveted joints are very strong providing they are correctly designed and assembled. The joint must be designed so that the rivet is in shear and not in tension. Consider the head of the rivet as only being strong enough to keep the rivet in place. You must consider a number of factors when selecting a rivet and making a riveted joint. These include the material used for the rivet as well as the shape of its head.

Test your knowledge 2.16

The cylinder head of a motor cycle engine is held on by studs. The studs are screwed into the cylinder casting at one end and the cylinder head is secured by nuts at the other end. The nuts must not vibrate loose. Which end of the stud should have a coarse thread? Which end of the stud should have a fine thread? Explain your answer.

Test your knowledge 2.17

State whether a rivet should be loaded in shear or in tension. Explain your answer.

Test your knowledge 2.18

State the main advantage of a screwed joint over a riveted joint.

Test your knowledge 2.19

List THREE factors you would need to consider when selecting a screwed fastening for a particular application.

Test your knowledge 2.20

List THREE factors you would need to consider when selecting a rivet for a particular application.

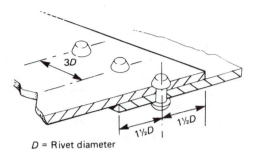

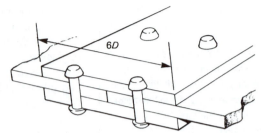

(a) Single-riveted lap joint

(e) Double-cover-plate butt joint

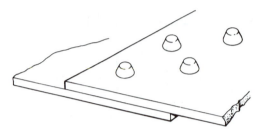

(b) Double-riveted lap joint

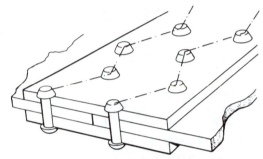

(f) Double-riveted, double-cover-plate butt joint, zigzag formation

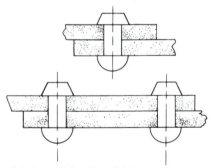

(c) Assembly of lap joints

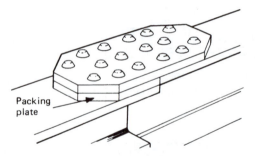

(g) Splice joint (horizontal)

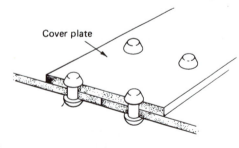

(d) Single-cover-plate butt joint

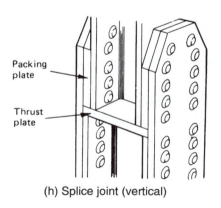

(h) Splice joint (vertical)

Figure 2.12 *Typical riveted joints*

The material from which the rivet is made must not react with the components being joined as this will cause corrosion and weakening. Also the rivet must be strong enough to resist the loads imposed upon it.

The rivet head chosen is always a compromise between strength and appearance. In the case of aircraft components, wind resistance must also be taken into account. Figure 2.13 shows some typical rivet heads and rivet types.

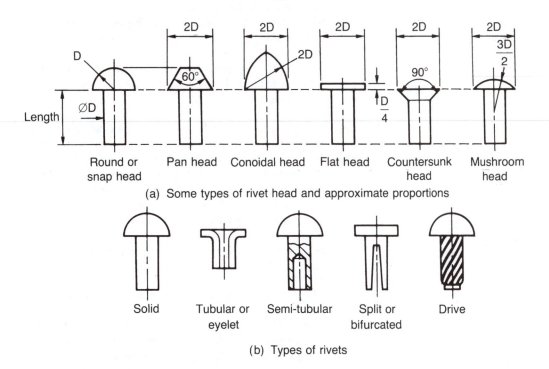

(a) Some types of rivet head and approximate proportions

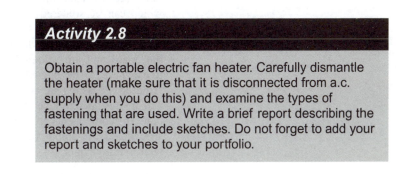

(b) Types of rivets

Figure 2.13 *Typical rivet heads and types*

2.4.10 Electrical and electronic components

When selecting electrical and electronic components, you have to consider the following factors:

- Has the component got the correct circuit value? For example, has it got the correct resistance or capacitance value. Also has it got the correct tolerance grade? The better the tolerance, the more expensive the component.

- Is it insulated to withstand the potential across the component and between the component and earth?
- If it is not enclosed, is it insulated against accidental contact (electric shock)?
- Can it pass the required current without overheating?
- Has it the correct power rating in watts?
- Will it fit onto the circuit board or chassis and is it suitable for connecting to the circuit board or associated components?
- If it is being used for telecommunications or data processing, is it suitable for the high frequencies used in these applications?

You have only to look through any electronics catalogue to see how many different types of electrical and electronic components are available. We only have room to look at a few items in this chapter. Large-scale manufacturers of electronic equipment would buy their components direct from the makers in bulk. For small-scale batch production and for prototype work it is usual to buy from a wholesale or retail supplier. This enables you to obtain all your requirements on a 'one-stop' purchasing basis.

Figure 2.14 shows some typical cables and hardware for electronic equipment:

(a) This shows examples of matrix board, strip board and a PCB. The matrix board (1) is a panel of laminated plastic perforated with a grid of holes. Pins can be fixed in the holes at convenient places for the attachment of such components as resistors and capacitors. They are merely attachment points and do not form part of the circuit. The strip board (2) is like a matrix board but is copper faced in strips on one side. The holes are the same pitch as the pins of integrated circuits which can be soldered into position. The components are placed on the insulated side of the board, with the copper strips underneath. The wire leads pass through the holes in the board and are soldered on the underside. The copper tracks have to be cut wherever a break in the circuit is required. PCBs (3) are custom made for a particular circuit and are designed to give the most efficient layout for the circuit. The components are installed in the same way as for the strip board. Assembly will be considered in greater detail later in this chapter.

(b) This is a typical flexible mains lead with PVC sheathing and colour coded PVC insulation. In selecting such a cable, the only factors you have to consider are its current handling capacity and its colour, providing the insulation is rated for mains use.

(c) This is a signal cable suitable for audio frequency analogue signals and for data processing signals. The conductors only need to have a limited current handling capacity, so many

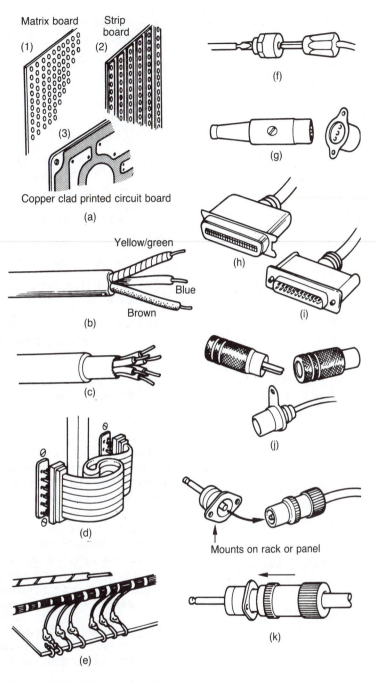

Figure 2.14 *Electronic wiring and connectors*

conductors can be carried in one cable. The conductors are surrounded by an earthed metal braid to prevent pick-up of external interference and corruption of the signal.

(d) This is a ribbon cable widely used in the manufacture of computers and the interconnection of computers and printers. Unlike the signal cable described earlier it is not screened against interference. However, it is cheaper and more easily terminated.

(e) Single cored PVC insulated wire is useful for making up wiring harnesses and for flying leads on PVC boards.

(f) This shows a 'banana' plug and socket. These are used with single cored, flexible conductors for low-voltage power supply connections.

(g) This shows a DIN type plug and socket. These are used in conjunction with multi-cored screened signal cables. They are available for 3-way to 8-way connections inclusive and are designed so that the plug can only be inserted into the socket in one position.

(h) This show a 36-way Centronics plug as widely used for making connections to the parallel port of a computer printer.

(i) This shows a 25-way D-type plug as widely used for making connections to the parallel output port of a computer.

(j) This shows a selection of phono type plugs and sockets. These are widely used for making signal lead connections to audio amplifiers.

(k) This shows a typical plug and socket. These are used at radio frequencies for connecting aerial leads to television sets.

Figure 2.15 shows a selection of electronic components that are widely used. Resistors are used to limit the flow of an electric current in a circuit. Carbon type resistors can be used on both direct current and alternating current circuits at any frequency since they are non-inductive. Wire wound resistors are inductive. They can only be used for direct current and mains frequency alternating current.

Capacitors are used to store electrical charges. Unlike batteries, they can be charged and discharged almost instantaneously. On the other hand, compared with batteries, they can only handle relatively small charges. Inductive devices such as chokes offer an impedance to alternating current. This increases as the frequency of the current gets higher. On direct current, inductive devices only offer the resistance of the wire from which they are wound.

(a) This shows a carbon rod resistor. They come in a wide range of resistance values and power ratings. These have quite wide tolerances but are suitable for general usage.

(b) This shows a high-stability resistor. These are precision resistors made from carbon film, metal film or oxide film. They are much more expensive than the ordinary carbon rod resistors. They are made to closer tolerances and are less susceptible to changes in resistance with changes in temperature. They are widely used in measuring instruments, computer applications and high-stability radio frequency oscillators.

(c) This shows a wire-wound, vitreous enamelled resistor. These are inductive and only suitable for direct current, and mains

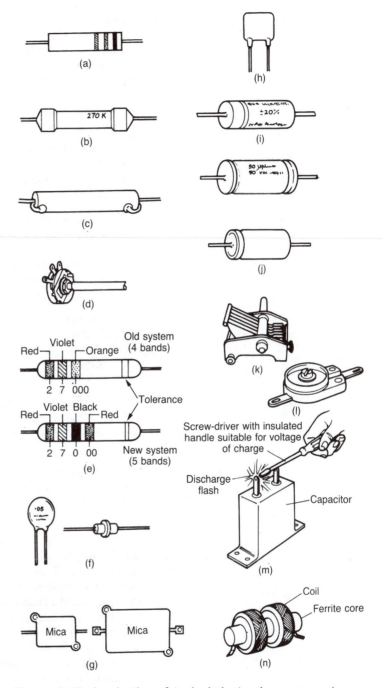

Figure 2.15 *A selection of typical electronic components*

frequency applications. They are used where high-power ratings are required.

(d) This shows a typical carbon track variable resistor. These are used for volume control and tone control circuits.

(e) Carbon rod and high-stability resistors are too small for much information to be marked on them so they are usually coded in some way. The colour code is as follows:

There are two ways of applying the colour code. Let us consider the old way first. This system had four colour bands. Three represented the resistance of the resistor and one represented the tolerance. Reading from the end of the resistor, the example in Figure 2.15(e) shows bands coloured red, violet, orange:

- The first band is the first number, red = 2
- The second band is the second number, violet = 7
- The third band is the number of zeros, orange = 3 or 000.

So the resistance of our resistor is 27,000 Ω (or 27 kΩ).

If the resistor has a silver band as its fourth band, its resistance could range from 10% below its nominal value to 10% above its nominal value. That is, from 24,300 Ω to 29,700 Ω. Such a wide range of values would be unacceptable for many applications, so additional close tolerance bands have been added; these are:

0.1% Violet
0.25% Blue
0.5% Green
1% Brown
2% Red

The new system uses five bands, four for the resistance value and one for the tolerance. The example in Figure 2.15(e) shows bands of red, purple, black, red:

- The first band is the first number, red = 2
- The second band is the second number, violet = 7
- The third band is the third number, black = 0
- The fourth band is the number of zeros, red = 2 or 00

So the resistance of our resistor is once again 27,000 Ω.

If the resistor has a gold band as its fifth band, its resistance could range from 5% below its nominal value to 5% above its nominal value. That is, from 25,650 to 28,350 Ω. The additional band provides for intermediate values of resistance. For example, if we had required 27,200 Ω ±1% the colours would have been red, violet, red, red, and a tolerance band coloured brown. This value could not have been achieved with the older system. Think about it.

A number and letter code is also used on circuit diagrams and often found printed on high-stability resistors. This is best explained by some examples:

0.47 Ω would be marked R47
4.7 Ω would be marked 4R7
47 Ω would be marked 47R
100 Ω would be marked R100
1 kΩ would be marked 1k0
10 kΩ would be marked 10k
47 MΩ would be marked 47M

Note that k = kilo = ×1000 and that M = mega = ×1,000,000.

(f) This shows examples of metallised ceramic capacitors. These are widely used in telecommunications equipment and in computers where high stability and compact size are required.

(g) This shows examples of silvered mica capacitors. These are also high-stability capacitors suitable for radio frequency tuned circuits and for pulse operation.

(h) This shows a moulded polyester capacitor. These are self-healing and are widely used on PCBs. They offer high values of capacitance in a small case size. They have a low inductance and low loss characteristics.

(i) This shows a polystyrene foil capacitor. These have ousted the foil and waxed paper capacitors found in some old equipment. They have low self-inductance, low high-frequency losses and a long life. They are used for signal coupling and filter circuits.

(j) This shows some electrolytic capacitors. These are used where very high values of capacitance are required: for example, smoothing capacitors in power supplies. Normally these are polarised and they can only be used in direct current circuits. They must always be connected into the circuit in the correct direction as indicated on the case. Bi-polar electrolytic capacitors are available for use with low-voltage alternating currents and as signal coupling capacitors in audio amplifiers.

(k) This shows a capacitor whose capacitance can be varied. Such a variable capacitor is wired in series or in parallel with an inductance (coil) to form a resonant (tuned) circuit. This is the way the tuning circuit of your radio works.

(l) This shows a preset pressure capacitor. The capacitance increases when the screw is rotated in a clockwise direction to tighten it. Such devices are used in pre-tuned circuits that do not have to be varied once they have been set.

(m) SAFETY: Large capacitors can store substantial charges of electricity at high voltages. Before handling such capacitors always discharge them as shown, either with a suitable screwdriver or with a length of insulated wire.

(n) This shows a typical inductor or coil. These may be air cored or they may be wound on a ferrite core to increase their inductance for a given size. Chokes have a single winding. Transformers may have a single winding with tappings (auto-transformer) or they may have two or more windings. Transformers can only be used on a.c. circuits, either power or signal circuits.

Figure 2.16 shows some solid state devices.

(a) Shows some thermistors. The resistance of these devices falls off rapidly as the temperature increases. They can be used as sensors to activate thermal protection devices. A common application is as sensors in car engines to activate the temperature gauge on the instrument cluster. As the water temperature

(a) Rod type
These special resistance elements have a very high negative temperature coefficient of resistance, making them suitable as protective elements in a wide range of circuits

Type	Resistance Cold	Resistance Hot	Dimensions in mm
TH-1A	650 Ω	37 Ω at 0.3 A	L. 38 Dia. 11
TH-2A	3.8 kΩ	44 Ω at 0.3 A	L. 32 Dia. 8
TH-3	370 Ω	28 Ω at 0.3 A	L. 22 Dia. 12
TH-5	4 Ω	0.4 Ω at 1 W (max)	Dia. 10 H. 4-5

(a)

(b) Bead type
Miniature glass-encapsulated thermistors for amplitude control and timing purposes (Types TH-B15, TH-B18) or temperature measurement (Types TH-B11, TH-B12).
Selection tolerance ±20% at 20°C

Type	Resistance at 20°C	Minimum resistance	Dimensions in mm
TH-B11	1 MΩ	170 Ω	L10 Dia 2.5
TH-B12	2 kΩ	115 Ω	
TH-B15	100 kΩ	320 Ω	L25 Dia. 4
TH-B18	5 kΩ	100 Ω	L.38 Dia. 10

(b)

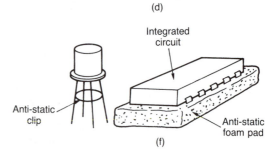

Figure 2.16 *A selection of thermistors and solid-state devices*

rises, the resistance of the sensor falls and the current through the circuit increases. This causes the temperature gauge to show a higher reading. Although measuring current, the scale of the instrument is calibrated in degrees of temperature.

(b) As with (a) above, (b) shows thermistors.

(c) This shows some diodes. These are electronic switching devices that only allow the current to flow in one direction. To ensure that they are connected in the circuit the correct way round, the positive end is either chamfered, or it has a red band. On larger, metal-cased diodes the diode symbol is printed on the case with the arrow head pointing to the positive pole of the diode.

(d) This shows a selection of transistors. These are switching devices that allow a small current to control a larger current. Small changes in the applied current can cause corresponding and amplified changes in the larger current. Power transistors have metal cases that can be bolted to a heatsink. If a transistor (or any other solid state device) overheats it is destroyed. For this reason heatsinks should be clamped to the legs of solid state devices whilst they are being soldered into the circuit. The shape of the transistor case gives an indication as to the correct way to connect it into the circuit. Consult manufacturers' data sheets for the different shapes and the corresponding connections.

(e) This shows a typical integrated circuit. These contain many components on one chip. Complete amplifiers, radio receivers, timers and many other devices can come ready packaged in a single case ready for installation on a circuit board.

(f) We have already mentioned the care that must be taken to prevent a solid-state device being destroyed by overheating. Equally important is the care that must be taken with some devices to prevent them being destroyed by electrostatic charges. Two ways of protecting such devices are shown in Figure 2.16(f). The anti-static clip must not be removed until the device has been installed in the circuit. Whilst installing such devices, the circuit board, the soldering iron and the installer must all be bonded to the same earth point. The installer wears a wrist clamp connected to earth via a flexible copper braid.

Test your knowledge 2.21

A resistor is marked with four coloured bands, as follows: Red, Red, Orange, Gold. What is the value and tolerance of this resistor?

Test your knowledge 2.22

Explain why large value capacitors can be dangerous.

Activity 2.9

Obtain a portable radio. Carefully dismantle the radio (make sure that it is disconnected from a.c. supply or remove the batteries when you do this) and examine the types of electronic component that are used in it. Write a brief report describing the main types of component that are used and include a sketch showing how they are laid out on the PCB. Do not forget to add your report and sketches to your portfolio.

2.5 Making engineered products

There are many stages in many aspects to making an engineered product including:

- the selection of appropriate materials (e.g. metals, plastics, composite materials)
- the selection of appropriate processes (e.g. those used for cutting, shaping, joining and finishing)

- the selection of appropriate tools and equipment (e.g. pillar drills, soldering irons, milling machines)
- the selection and application of appropriate quality assurance measures.

We will start out by describing the techniques that we use for measuring components and parts.

2.5.1 Measurement

Before you can mark out a component or check it during manufacture you need to know about engineering measurement. All engineering measurements are comparative processes. You compare the size of the feature to be measured with a known standard. Figure 2.17 shows a

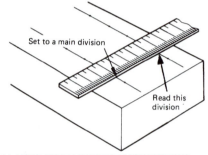

(a) MEASURING THE DISTANCE BETWEEN TWO SCRIBED LINES

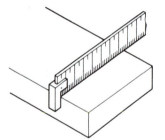

(b) MEASURING THE DISTANCE BETWEEN TWO FACES USING A HOOK RULE

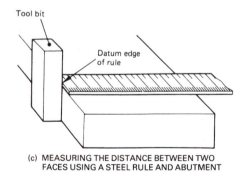

(c) MEASURING THE DISTANCE BETWEEN TWO FACES USING A STEEL RULE AND ABUTMENT

Figure 2.17 *Using a steel rule*

steel rule and how to use it. The distance between the lines or the width of the work is being compared with the rule. In this instance the rule is our standard of length. The rule should be made from spring steel and the markings should be engraved into the surface of the rule. The edges of the rule should be ground so that it can be used as a straight edge, and the datum end of the rule should be protected from damage so that the accuracy of the rule is not lost. NEVER use a rule as a screwdriver or for cleaning out the T-slots on machine tools.

To increase the usefulness of the steel rule and to improve the accuracy of taking measurements, accessories called calipers are used. These are used to transfer the distances between the faces of the work to the distances between the lines engraved on the rule. Figure 2.18 shows some different types of inside and outside calipers. It also shows how to use calipers.

A steel rule can only be read to an accuracy of about ± 0.5 mm. This is rarely accurate enough for precision engineering purposes. Figure 2.19(a) shows a vernier caliper and how it can take inside and outside measurements. Typical vernier scales are shown in Figure 2.19(b). Some verniers have scales that are different to the ones shown in

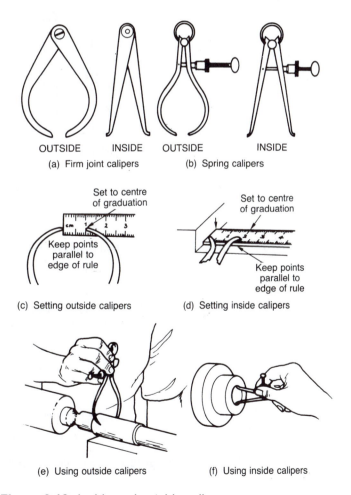

Figure 2.18 *Inside and outside calipers*

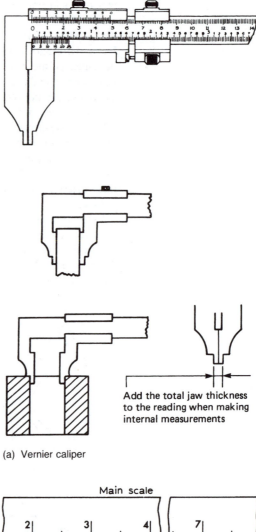

Add the total jaw thickness to the reading when making internal measurements

(a) Vernier caliper

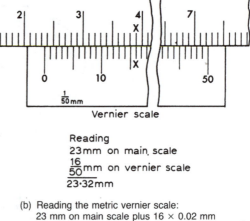

Main scale

Vernier scale

Reading
23mm on main scale
$\frac{16}{50}$mm on vernier scale
23·32mm

(b) Reading the metric vernier scale:
23 mm on main scale plus 16 × 0.02 mm
on vernier scale gives total 23.32 mm

Figure 2.19 *Vernier calipers*

Figure 2.19(b). Always check the scales before taking a reading. Like all measuring instruments, a vernier caliper must be treated carefully and it must be cleaned and returned to its case whenever it is not in use.

Vernier calipers are difficult to read accurately even if you have good eyesight. A magnifying glass is helpful. The larger sizes of vernier caliper are quite heavy and it is difficult to get a correct and consistent 'feel' between the instrument and the work. An alternative instrument is the micrometer caliper. Figure 2.20 shows a micrometer caliper and the method of reading its scales.

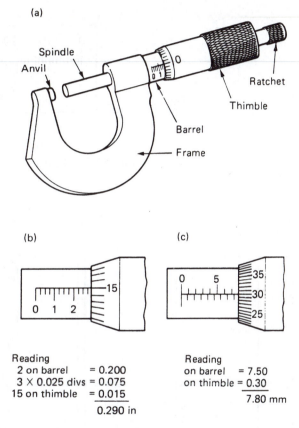

Figure 2.20 *Micrometer caliper*

Test your knowledge 2.23

State THREE precautions that you should take to keep a steel rule in good condition.

Test your knowledge 2.24

Explain how you would obtain a constant measuring pressure when using a micrometer caliper.

Micrometer calipers are more compact than vernier calipers. They are also easier to use. The ratchet on the end of the thimble ensures that the contact pressure is kept constant and at the correct value. Unfortunately, micrometer calipers have only a limited measuring range (25 mm), so you need a range of micrometers moving up in size in 25 mm steps (0–25, 25–50, 50–75 mm and so on to the largest size.) You also need a range of inside micrometers, and depth micrometers.

Angular measurements
The most frequently measured angle is 90°. This is a right-angle and surfaces at right-angles to each other are said to be perpendicular.

Test your knowledge 2.25
(a) Determine the reading of the micrometer caliper scales shown in Figure 2.21(a).
(b) Determine the reading of the vernier caliper scales shown in Figure 2.21(b).

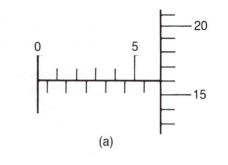

(a)

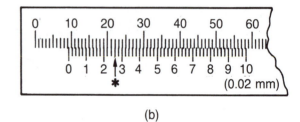

(b)

Figure 2.21 *See Test your knowledge 2.25*

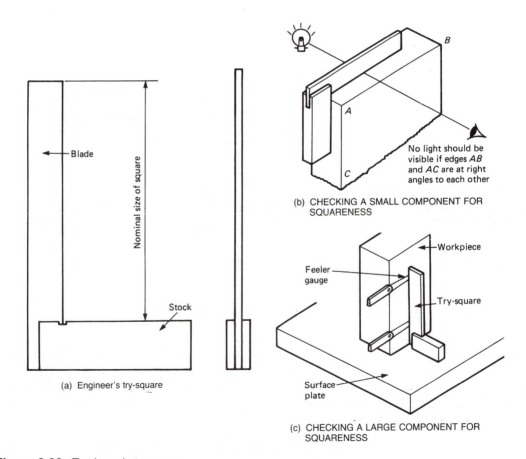

No light should be visible if edges *AB* and *AC* are at right angles to each other

(b) CHECKING A SMALL COMPONENT FOR SQUARENESS

Workpiece

Feeler gauge

Try-square

Surface plate

(c) CHECKING A LARGE COMPONENT FOR SQUARENESS

Blade

Nominal size of square

Stock

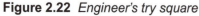

(a) Engineer's try-square

Figure 2.22 *Engineer's try square*

Right angles are checked with some sort of try square. Figure 2.22 shows a typical engineer's try square and two ways in which it can be used.

For angles other than a right-angle, a protractor is used. This may be a simple plain protractor as shown in Figure 2.23(a) or it may have a vernier scale (vernier protractor) as shown in Figure 2.23(b).

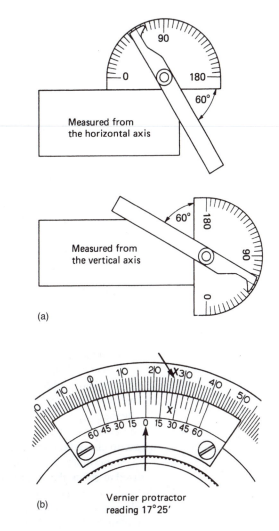

Figure 2.23 *Using a protractor*

Tolerancing and gauging

So far we have only considered measurement of size. This is usual when making a small number of components. However for quantity production it requires too high a skill level, is too time consuming and, therefore, too expensive. Since no product can be made to an exact size, nor can it be measured exactly, the designer usually gives each dimension an upper and lower size. This is shown in Figure 2.24. If the component lies anywhere between the upper and lower limits of

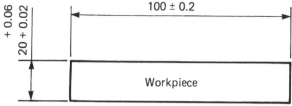

Dimensions in millimetres

EXAMPLE 1

Nominal size	100 mm
Limits (low)	99.8 mm
Limits (high)	100.2 mm
Tolerance	0.4 mm
Deviation	± 0.2 mm
Mean size	100.0 mm

EXAMPLE 2

Nominal size	20 mm
Limits (low)	20.2 mm
Limits (high)	20.6 mm
Tolerance	0.4 mm
Deviation	+0.02, +0.06
Mean size	20.4 mm

Figure 2.24 *Use of tolerances*

size it will function correctly. The closer the limits, the more accurately the component will work, but the more expensive it will be to make.

A major advantage of using tolerance dimensions is that they can be checked without having to be measured. Gauges can be used instead of measuring instruments. This is easier, quicker and much cheaper. Figure 2.25 shows how a caliper gauge can be used to check the thickness of a component. Plug gauges are used in a similar manner to check hole sizes.

Some more gauges are shown in Figure 2.26. Radius gauges are used to check the corner radii of components. Feeler gauges are used to check the gap between components: for example, the valve tappet clearances in a motor vehicle engine. Thread gauges are used to check the pitch of screw threads.

Activity 2.10

Select suitable measuring instruments and explain how you would check the dimensions for the component shown in Figure 2.28. Present your work in the form of a brief written report including sketches where appropriate. Do not forget to include your report and sketches in your portfolio.

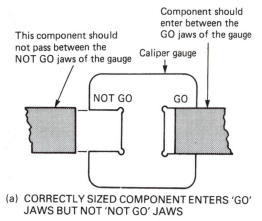

Component should
enter between the
GO jaws of the gauge

This component should
not pass between the
NOT GO jaws of the gauge

Caliper gauge

NOT GO   GO

(a) CORRECTLY SIZED COMPONENT ENTERS 'GO'
JAWS BUT NOT 'NOT GO' JAWS

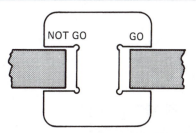

NOT GO   GO

(b) UNDERSIZE COMPONENT ENTERS 'GO' AND
'NOT GO'

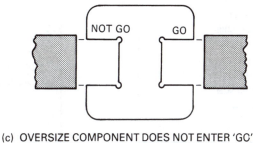

NOT GO   GO

(c) OVERSIZE COMPONENT DOES NOT ENTER 'GO'
OR 'NOT GO'

Figure 2.25 *Using a caliper gauge to check the thickness of
a component*

2.5.2 *Marking out*

The reasons why you mark out components before making them are
as follows:

- To provide you with guide lines to work to. Where only limited
  accuracy is required, marking out also controls the size and
  shape of the workpiece and the position of any holes.

- As a guide to a machinist when setting up and cutting. In this
  instance the final dimensional control comes from the use of
  precision measuring instruments together with the micrometer
  dials on the machine controls.

Key point

Marking out is performed before a
metal is cut or machined.

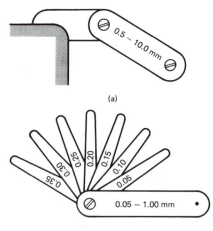

(a)

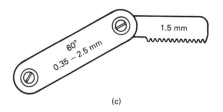

(b)

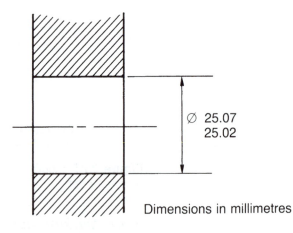

(c)

Figure 2.26 *Radius and feeler gauges*

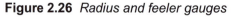

∅ 25.07
25.02

Dimensions in millimetres

Figure 2.27 *See Test your knowledge 2.29 and 2.30*

- Marking out also indicates if sufficient machining allowance has been left on cast or forged components. It also indicates whether or not such features as webs, flanges and cored holes have been correctly positioned.

Scribed lines and centre marks
Scribed lines are fine lines cut into the surface of the material being marked out by the point of a scribing tool (scriber). An example of a

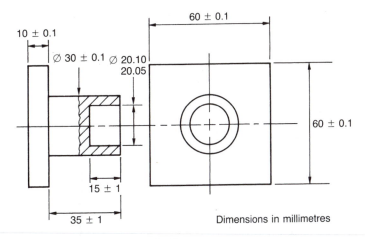

Figure 2.28 *See Activity 2.10*

Test your knowledge 2.31

Write down the readings for:

(a) the micrometer shown in Figure 2.29(a)
(b) the vernier shown in Figure 2.29(b).

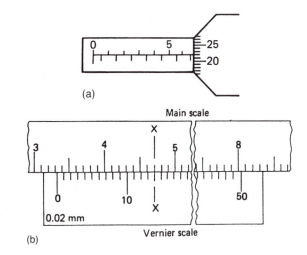

Figure 2.29 *See Test your knowledge 2.31*

scriber is shown in Figure 2.30(a). To ensure that the scribed line shows up clearly, the surface of the material to be marked out is coated with a thin film of a contrasting colour. For example, the surfaces of the casting to be machined are often whitewashed. Bright metal surfaces can be treated with a marking out 'ink'. Plain carbon steels can be treated with copper sulphate solution which copper plates the surface of the metal. This has the advantages of permanence. Care must be taken in its use as it will attack any marking out and measuring instruments into which the copper sulphate comes into contact.

Centre marks are made with a dot punch as shown in Figure 2.30(b) or with a centre punch as shown in Figure 2.30(c). A dot punch has a fine conical point with an included angle of about 60°. A centre punch is heavier and has a less acute point angle of about 90°. It is used for making a centre mark for locating the point of a twist drill and

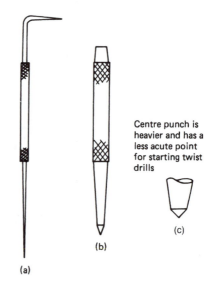

Centre punch is heavier and has a less acute point for starting twist drills

(a)   (b)   (c)

Figure 2.30 *Scriber and centre punch*

preventing the point from wandering at the start of a cut. The dot punch is used for two purposes when marking out:

- A scribed line can be protected by a series of centre marks made along the line as shown in Figure 2.31(a). If the line is accidentally removed, it can be replaced by joining up the centre marks. Further, when machining to a line as shown in Figure 2.31(b), the half-marks left behind are a witness that the machinist has 'split the line'.

- Secondly, dot punch marks are used to prevent the centre point of dividers from slipping when scribing circles and arcs of circles as shown in Figure 2.31(c). The correct way to set divider points is shown in Figure 2.31(d).

When a centre punch is driven into the work, distortion can occur. This can be a burr raised around the punch mark, swelling of the edge of a component, or the buckling of thin material.

2.5.3 Equipment for marking out

From what we have already seen, your basic requirements for marking out are:

- A scriber to produce a line
- A rule to measure distances and act as a straight edge to guide the point of the scriber
- Dividers to scribe circles and arcs of circles as shown in Figure 2.31(c).

In addition, you require hermaphrodite (odd-leg) calipers as shown in Figure 2.32(a) and a try square. Odd-leg calipers are used to scribe lines parallel to a datum edge. A try square and scriber are used to scribe a line at right-angles to a datum edge as shown in Figure 2.32(b).

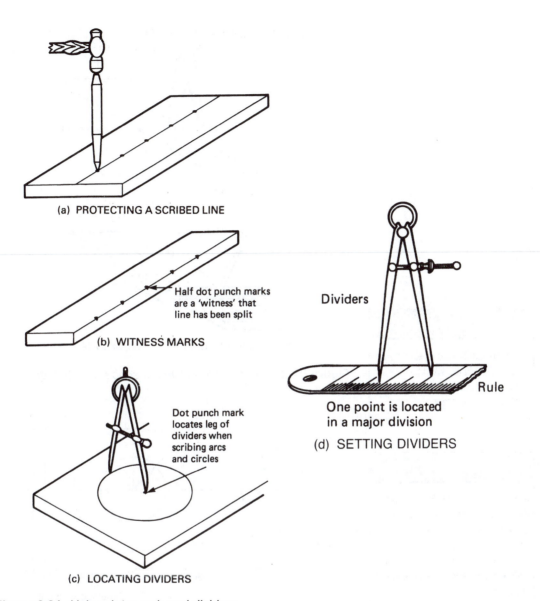

(a) PROTECTING A SCRIBED LINE

Half dot punch marks
are a 'witness' that
line has been split

(b) WITNESS MARKS

Dividers

Rule

One point is located
in a major division

(d) SETTING DIVIDERS

Dot punch mark
locates leg of
dividers when
scribing arcs
and circles

(c) LOCATING DIVIDERS

Figure 2.31 *Using dot punch and dividers*

Alternatively, lines can be scribed parallel to a datum edge using a surface table and scribing block as shown in Figure 2.34(a). In this example the scribing point is set to a steel rule, so the accuracy is limited. Alternatively, the line can be scribed using a vernier height gauge as shown in Figure 2.34(b). This is very much more accurate. Where extreme accuracy is required, slip gauges and slip gauge accessories can be used as shown in Figure 2.34(c).

Cylindrical components are difficult to mark out since they tend to roll about. To prevent this they can be supported on vee-blocks as shown in Figure 2.35(a). Vee-blocks are always made and sold in boxed sets of two. In order that the axis of the work is parallel to the surface plate or table, you must always use such a matched pair of

Key point

Tools used for marking out include scribers, rules, dividers and calipers.

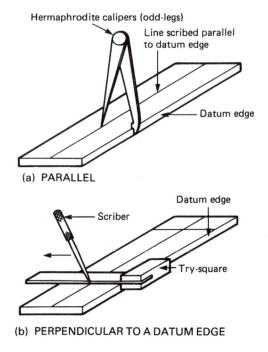

(a) PARALLEL

(b) PERPENDICULAR TO A DATUM EDGE

Figure 2.32 *Using odd-leg calipers, scriber and try square*

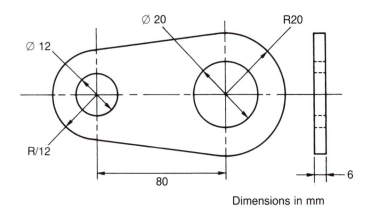

Dimensions in mm

Figure 2.33 *See Test your knowledge 2.32 and 2.33*

vee-blocks, and make sure that, after use, they are always put away as a pair.

Another useful device for use on cylindrical work is the box square shown in Figure 2.35(b). This is used for scribing lines along cylindrical work parallel to the axis. A centre finder is shown in Figure 2.35(c). This is used to scribe lines that pass through the centre of circular blanks or the ends of cylindrical components.

Datum points, lines and surfaces
First we had better revise what we know about rectangular and polar coordinates. Examples of these are shown in Figure 2.36.

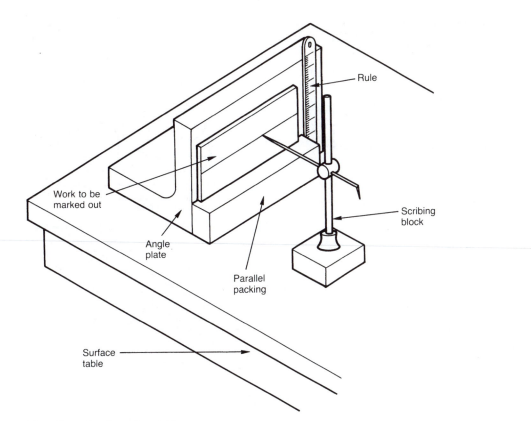

(a) marking out on the surface table.

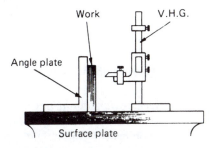

(b) Marking out with the vernier height

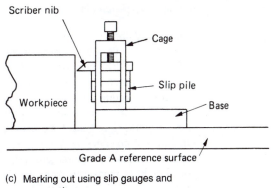

(c) Marking out using slip gauges and
accessories

Figure 2.34 *Marking out using a surface table*

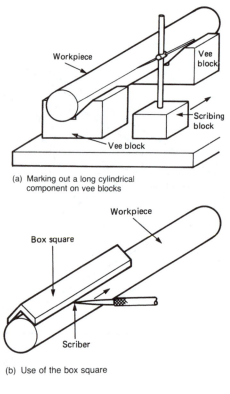

(a) Marking out a long cylindrical component on vee blocks

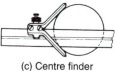

(b) Use of the box square

(c) Centre finder

Figure 2.35 *Marking cylindrical components*

Rectangular coordinates

The point A in Figure 2.36(a) is positioned by a pair of ordinates (coordinates) lying at right-angles to each other. They also lie at right-angles to the datum edges from which they are measured. This system of measurement requires the production of two datum surfaces or edges at right-angles to each other. That is, two datum edges that are mutually perpendicular.

Polar coordinates

Polar coordinates consist of one linear distance and an angle as shown in Figure 2.36(b). Dimensioning in this way is useful when the work is to be machined on a rotary table. It is widely used when setting out hole centres round a pitch circle. Quite frequently both systems are used at the same time as shown in Figure 2.36(c). During our discussion on marking out, we have kept referring to datum points, datum lines and datum edges. A datum is any point, line or surface that can be used as a basis for measurement. When you go for a medical check-up, your height is measured from the floor on which you are standing. In this example the floor is the basis of measurement, it is the datum surface from which your height is measured.

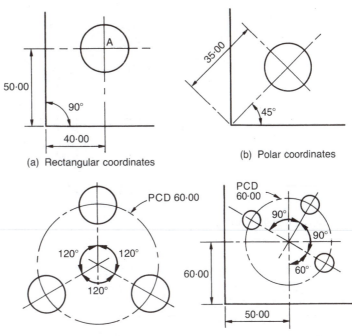

(a) Rectangular coordinates

(b) Polar coordinates

(c) Polar coordinates applied
to holes on a pitch circle

(d) Combined coordinates
Dimensions in mm

(Note: PCD = Pitch circle diameter)

Figure 2.36 *Datum points and coordinates*

Point datum
This is a single point from which a number of features are marked
out. For example, Figure 2.37(a) shows two concentric circles representing the inside and outside diameter of a pipe-flange ring. It also
shows the pitch circle around which the bolt hole centres are marked
off. All these are marked out using dividers or trammels (beam compasses) from a single-point datum.

Line datum
Any line from which, or along which a number of features are marked
out. An example of the use of line datums is shown in Figure 2.37(b).

Surface datum
This is also known as an edge datum and a service edge. It is the most
widely used datum for marking out solid objects. Two edges are

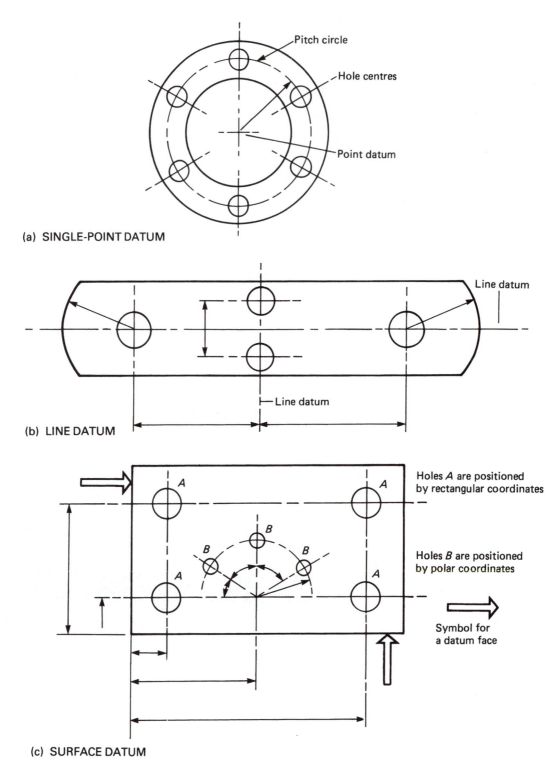

(a) SINGLE-POINT DATUM

(b) LINE DATUM

(c) SURFACE DATUM

Figure 2.37 *Point, line and surface datums*

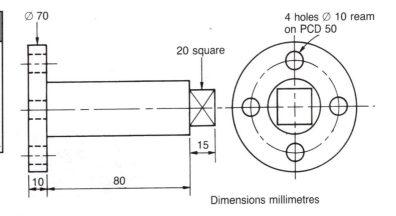

Figure 2.38 *See Activity 2.11*

accurately machined at right angles to each other and all the dimensions are taken from these edges. Figure 2.37(c) shows how dimensions are taken from surface datums. It shows both rectangular and polar coordinates. Alternatively, the work can be clamped to an angle-plate. The mutually perpendicular edges of the angle-plate provide the surface datums. In this instance there is no need to machine the edges of the workpiece at right-angles.

This concludes our work on measurement and marking out. You should now be ready to put some of this knowledge into good use in the next section when we shall be introducing some basic hand tools and processes.

2.5.4 Hand tools and benchwork

The next section introduces you to a variety of hand tools and how they are used. We will start with the fitter's bench and vice that is used for holding work whilst performing cutting, chiselling or filing.

Fitter's bench and vice

A fitter's bench should be substantial and rigid. This is essential if accurate work is to be performed on it. It should be positioned so that it is well lit by both natural and artificial light without glare or shadows. It should be equipped with a fitter's vice. A plain screw vice is shown in Figure 2.39(a) and a quick action vice is shown in Figure 2.39(b). In the latter type of vice, the jaws can be quickly pulled apart when the lever at the side of the screw handle is released. The screw is used for closing the jaws and clamping the work in the usual way.

The jaws of a fitter's vice are serrated and hardened to prevent the work from slipping. This also marks the surfaces of the work. For fine work with finished surfaces, the serrated jaws should be replaced with hardened and ground smooth jaws. Alternatively vice shoes can be used. These are faced with a fibre compound and can be slipped over the serrated jaws when required. A pair of typical vice shoes are shown in Figure 2.39(c).

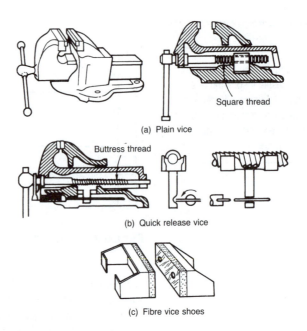

(a) Plain vice

Square thread

Buttress thread

(b) Quick release vice

(c) Fibre vice shoes

Figure 2.39 *A fitter's vice*

Cutting tools

Before we can discuss the cutting tools we use for bench fitting we need to look at the way metal is cut. Here are the basic facts.

Wedge angle

If you look at a hacksaw blade as shown in Figure 2.40(a), you can see that the teeth are wedge shaped. Figure 2.40(b) shows how the wedge angle increases as the material gets harder. This strengthens the cutting edge and increases the life of the tool. At the same time it reduces its ability to cut. Try cutting a slice of bread with a cold chisel!

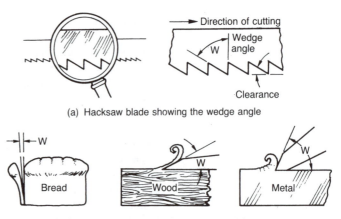

Direction of cutting

Wedge angle

W

Clearance

(a) Hacksaw blade showing the wedge angle

W

Bread

W

Wood

W

Metal

(b) Wedge angles for various materials

Figure 2.40 *Hacksaw blade and wedge angles*

Test your knowledge 2.36

Calculate the wedge angle for a tool if the clearance angle is 5° and the rake angle is 17°.

Test your knowledge 2.37

Explain why cutting fluids (suds) do not have to be used when filing, but are used when turning.

Test your knowledge 2.38

Explain why metal cutting tools have a larger wedge angle than wood cutting tools.

Test your knowledge 2.39

Explain why metal cutting tools need a clearance angle.

Test your knowledge 2.40

Sketch an example of any lathe tool that cuts orthogonally and any lathe tool that cuts obliquely.

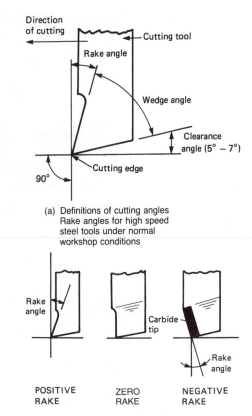

(a) Definitions of cutting angles
   Rake angles for high speed steel tools under normal workshop conditions

POSITIVE RAKE   ZERO RAKE   NEGATIVE RAKE

(b) Comparison of rake angles

Figure 2.41 *Rake angle*

Clearance angle
If you look at the hacksaw blade in Figure 2.40(a), you can see that there is a clearance angle behind the cutting edge of the tooth. This is to enable the tooth to cut into the work.

Rake angle
This angle controls the cutting action of the tool (Table 2.7). It is shown in Figure 2.41. I hope you can see that the wedge angle, clearance angle and rake angle always add up to 90°. This is true even when the rake angle is zero or negative as shown in Figure 2.41(b). The greater the rake angle the more easily the tool will cut. Unfortunately, the greater the rake angle, the smaller the wedge angle will be and the weaker the tool will be. Therefore, the wedge and rake angles have to be a compromise between ease of cutting and tool strength and life. The clearance angle remains constant at between 5° and 7°.

Orthogonal and oblique cutting
Figure 2.42(a) is a pictorial representation of the single point cutting tool shown in Figure 2.41. Notice how the cutting edge is at right angles to the direction in which the tool is travelling along the work. This is called orthogonal cutting. Now look at Figure 2.42(b). Notice how the cutting edge is inclined at an angle to the direction of cut.

Material	Rake angle
Aluminium alloy	30°
Brass (ductile)	14°
Brass (free-cutting)	0°
Cast iron	0°
Copper	20°
Phosphor bronze	8°
Mild steel	25°
Medium carbon steel	15°

Table 2.7 *Rake angle for various materials*

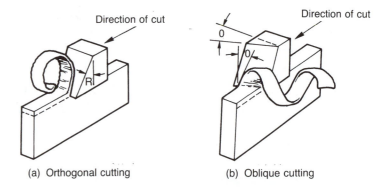

(a) Orthogonal cutting   (b) Oblique cutting

Figure 2.42 *Orthogonal and oblique cutting*

This is called oblique cutting. Oblique cutting results in a better finish than orthogonal cutting, mainly because the chip is thinner for a given rate of metal removal. This reduced thickness and the geometry of the tool allows the chip to coil up easily in a spiral.

Apart from threading operations, it is very rare to use a coolant or lubricant when using hand tools. However, the conditions are very different when using machine tools. Large amounts of metal are removed quickly, considerable energy is used to do this, and this energy is largely converted into heat at the cutting zone. The rapid temperature rise of the work and the cutting tool can lead to inaccuracy and short tool life. A *coolant* is required to prevent this. The chip flowing over the rake face of the tool results in wear. A lubricant is required to prevent this. Usually coolants are poor lubricants, and lubricants are poor coolants.

For general machining an emulsion of cutting oil (which also contains an *emulsifier*) and water is used. This has a milky-white appearance and is commonly known as 'suds'. On no account try to use a mineral lubricating oil. This cannot stand up to the temperatures and pressures found in the cutting zone. It is completely useless as a cutting lubricant or as a coolant. It gives off clouds of noxious fumes and it is a fire risk.

Cold chisels
Let us now see how the cutting angles discussed in the previous section can be applied to the basic bench tools.

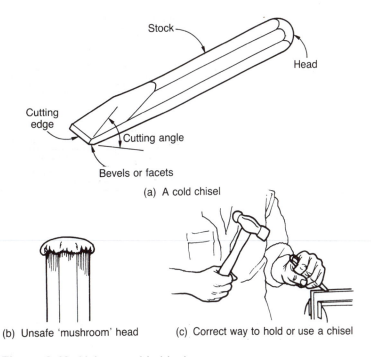

(a) A cold chisel

(b) Unsafe 'mushroom' head

(c) Correct way to hold or use a chisel

Figure 2.43 *Using a cold chisel*

Figure 2.43(a) shows a typical cold chisel and names its more important features. Constantly hitting the head of a chisel causes it to mushroom as shown in Figure 2.43(b). Never use a chisel with a mushroom head because bits of metal can fly off it when it is hit. These can cause an accident. When the mushroom head starts to form, it must be trimmed off on a grinding machine. Figure 2.43(c) shows the correct way to hold and use a cold chisel.

Safety when chipping
There are several important safety points that should always be observed when using a cold chisel:

- Never chip towards another person.
- Always chip towards a chipping screen.
- Always wear goggles when chipping.
- Always grind the mushroom head off a chisel before using it and make sure the cutting edge is sharp and in good condition. Regrind if necessary.

The chisel shown in Figure 2.43 is only one of many different types of chisel. Some further examples and their applications are shown in Figure 2.44. In the course of a working lifetime most fitters will make up many small chisels for special jobs. These are often made from hardened and tempered silver steel rod. In addition, there are the fine engraving chisels or 'gravers' used by die-sinkers and engravers.

The application of the cutting angles we discussed earlier can be applied to a chisel as shown in Figure 2.45. The point angle (wedge angle) and the angle of inclination are only a guide. In practice, a

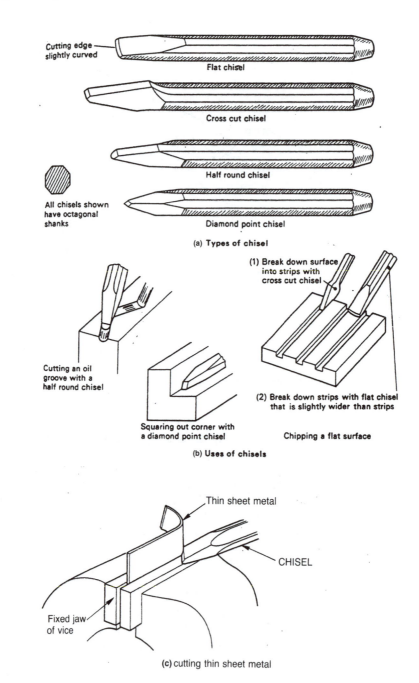

Figure 2.44 *Types of chisel and applications*

fitter does not work out the angle of inclination or the rake and clearance angle, but uses experience and the feel of the chisel as it cuts through the metal to present the chisel to the work at the correct angle.

Files are the most widely used and important tools for the fitter. The main parts of a file are named in Figure 2.46(a). Files are forged to shape from 1.2% plain carbon steel. After forging, the teeth are

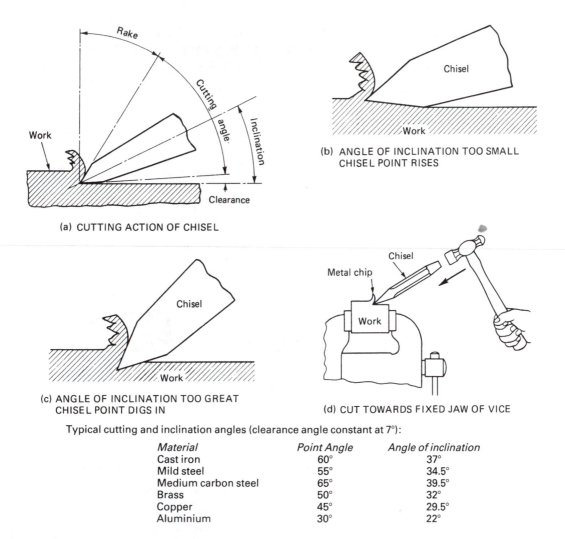

Typical cutting and inclination angles (clearance angle constant at 7°):

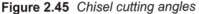

Material	Point Angle	Angle of inclination
Cast iron	60°	37°
Mild steel	55°	34.5°
Medium carbon steel	65°	39.5°
Brass	50°	32°
Copper	45°	29.5°
Aluminium	30°	22°

Figure 2.45 *Chisel cutting angles*

machine cut by a chisel-shaped tool as shown in Figure 2.46(b). The teeth of a single-cut are wedge shaped with the rake and clearance angles essential for metal cutting. Most files used in general engineering are double-cut. That is they have two rows of cuts at an angle to each other as shown in Figure 2.46(c).

Files are classified by the following features:

- length
- kind of cut
- grade of cut (roughness)
- profile
- cross-sectional shape or most common use.

The grades of cut are rough, bastard, second, smooth and dead smooth. These cuts vary with the length of a file. For example a short, second cut file will be smoother than a longer, smooth file. For further information on file cuts (see Table 2.8). The profiles and cross-sectional shapes of some typical files are shown in Figure 2.47.

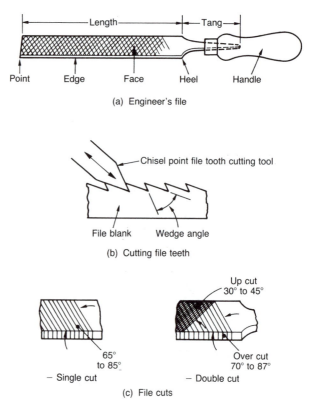

(a) Engineer's file

(b) Cutting file teeth

(c) File cuts

Figure 2.46 *An engineer's file*

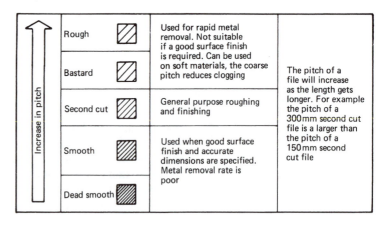

Table 2.8 *Types of file*

Figure 2.48 shows how a file should be held and used. To file flat is very difficult and the skill only comes with years of continual practice. *Cross-filing* is used for rapid material removal. *Draw filing* is only a finishing operation to improve the surface finish. It removes less metal per stroke than cross-filing and can produce a hollow surface, unless care is taken.

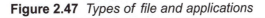

Type of file		Applications
⬛ ▭▭▭▭▭▭▭▭	Square	Filing of keyways and slots
△ ▭▭▭▭▭▭	Three square	Filing of angled surfaces
▽ ▭▭▭▭▭▭	Knife	Filing of acute angles
▯ ▭▭▭▭▭▭	Hand	These two files are the general-purpose tools for filing flat surfaces and convex profiles
▯ ▭▭▭▭▭▭	Flat	
⚙ ▭▭▭▭▭▭	Round	Used for enlarging or elongating holes
🔵 ▭▭▭▭▭▭	Half round	Filing of concave profiles

Figure 2.47 *Types of file and applications*

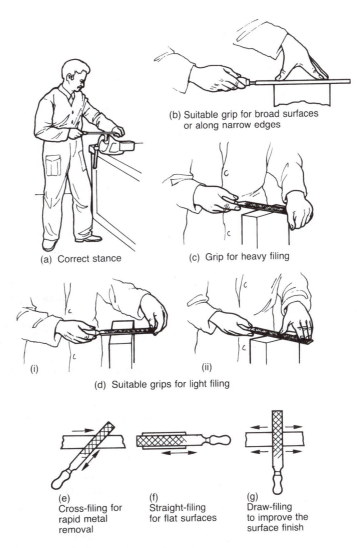

(a) Correct stance

(b) Suitable grip for broad surfaces or along narrow edges

(c) Grip for heavy filing

(i)   (ii)

(d) Suitable grips for light filing

(e) Cross-filing for rapid metal removal

(f) Straight-filing for flat surfaces

(g) Draw-filing to improve the surface finish

Figure 2.48 *Correct use of a file*

The spaces between the teeth of a file tend to become clogged with bits of metal. This happens mostly when filing soft metals. It is called *pinning*. The clogged teeth tend to leave heavy score marks in the surface of the work. These marks are difficult to remove. The file should be kept clean and a little chalk should be rubbed into the teeth to prevent pinning. Files are cleaned with a file brush called a *file card*.

Hacksaws and sawing

A typical hacksaw frame and blade is shown in Figure 2.49(a). The frame is adjustable so that it can be used with blades of various lengths. It is also designed to hold the blade in tension when the wing nut is tightened. The blade is put into the frame so that the teeth cut on the forward stroke. Figure 2.49(b) shows how a hacksaw should be held when being used.

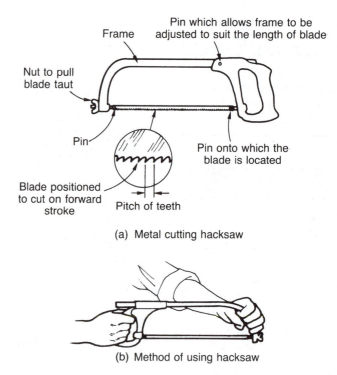

(a) Metal cutting hacksaw

(b) Method of using hacksaw

Figure 2.49 *A typical hacksaw frame and blade*

There are a variety of blade types available:

- High-speed steel 'all hard' blades are the most rigid and give the most accurate cut. However, they are brittle and easily broken when used by an inexperienced person.
- High-speed steel 'soft back' blades have a good life and, being more flexible, are less easily broken.
- Carbon steel flexible blades are satisfactory for occasional use on soft non-ferrous metals. They are cheap and not easily broken. Unfortunately, they only have a limited life when cutting steels.

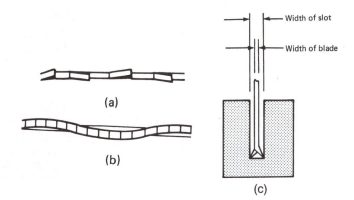

Figure 2.50 *The 'set' of a hacksaw blade*

To prevent the blade from jamming in the slot it makes as it cuts, all saw blades are given a 'set'. This is shown in Figure 2.50. Coarse pitch hacksaw blades and power-saw blades have the individual teeth set to the left and to the right with either the intermediate teeth or every third tooth left straight to clear the slot. This is shown in Figure 2.50(a). This is not possible with fine pitch blades, and the blade as a whole is given a 'wave' set as shown in Figure 2.50(b). The effect of set on the cut being made is shown in Figure 2.50(c). If you have to change a blade part way through a cut, never continue in the old slot. As the set of the old blade will have worn, the new blade will jam in the old cut and break. Always start a new cut to the side of the failed cut.

The sizes of hacksaw blades are now given in metric sizes. The length (between the fixing hole centres), the width and the thickness. However, the pitch of the teeth is still given as so many teeth per inch.

The fewer teeth per inch the coarser will be the cut, the more quickly will the metal be removed, and the greater will be the set so that there is less chance of the blade jamming. However, there should always be at least three teeth in contact with the work at any one time. Therefore, the thinner the metal being cut the finer the pitch of the blade that should be used. Some typical examples are given in Table 2.9.

Teeth per inch	Material to be cut	Blade applications
32	Up to 3 mm	Thin sheets and tubes Hard and soft materials (thin sections)
24	3–6 mm	Thicker sheets and tubes Hard and soft materials (thicker sections)
18	6–12 mm	Heavier sections such as mild steel, cast iron, aluminium, brass, copper, bronze
14	Greater than 12 mm	Soft materials (such as aluminium brass, copper, bronze) with thicker sections

Table 2.9 *Hacksaw blade sizes and applications*

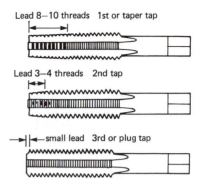

Figure 2.51 *Hand taps*

Screw thread cutting

Internal screw threads are cut with taps. A set of straight fluted hand taps are shown in Figure 2.51. The difference between them is the length of the lead. The taper tap should be used first to start the thread. Great care must be taken to ensure that the tap is upright in the hole and it should be checked with a try square. The second tap is used to increase the length of thread and can be used for finishing if the tap passes through the work. The third tap is used for *bottoming* in blind holes.

The hole to be threaded is called a *tapping size* hole and it is the same size or only very slightly larger than the core diameter of the thread. Drill diameters for drilling tapping size holes for different screw threads can be found in sets of workshop tables. For example, the tapping size drill for an M10 × 1.5 thread is 8.5 mm diameter.

A *tap wrench* is used to rotate the taps. There are a variety of different styles available depending upon the size of the taps. An example of a suitable wrench for small taps is shown in Figure 2.52. Taps are very fragile and are easily broken, particularly in the small sizes. Once a tap has been broken into a hole, it is virtually impossible to get it out without damaging or destroying the workpiece.

Figure 2.52 *A tap wrench*

Taps are relatively expensive and should be looked after carefully. High-speed steel ground thread taps are the most expensive. However, they cut very accurate threads and, with careful use, last a long time. Carbon steel cut thread taps are less accurate and less expensive and have a reasonable life when cutting the softer non-ferrous metals. Whichever sort of taps are used, they should always be well lubricated. Traditionally tallow was used, but nowadays proprietary screw-cutting lubricants are available that are more effective.

External threads are cut using split button dies in a die holder as shown in Figure 2.53. One face of the die is always marked up with details of the thread and the maker's logo. This should be visible when the die is in the die holder. Then the lead is on the correct side for

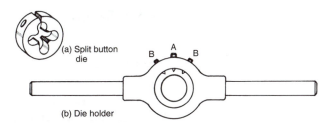

Figure 2.53 *A die holder*

starting the cut. Screw A is used to spread the die for the first cut. The screws marked B are used to close the die until it gives the correct finishing cut. This is judged by using a standard nut or a screw thread gauge. The nut or gauge should run up the thread without binding or without undue looseness.

Again, the die must be started square with the workpiece or a 'drunken' thread will result. Also, a thread cutting lubricant should be used. Like thread cutting taps, dies are available in carbon steel cut thread and high-speed steel ground thread types. For both taps and dies, each set only cuts one size and pitch of thread and one thread form.

Spanners and keys
In addition to cutting tools a fitter should also have a selection of spanners and keys available for dismantling and assembly purposes. Figure 2.54 shows a selection of spanners and keys. These are carefully proportioned so that a person of average strength will be able to tighten a screwed fastening correctly.

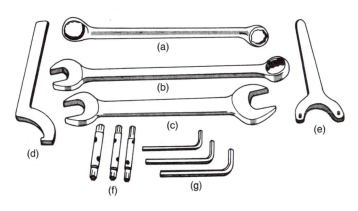

Figure 2.54 *A selection of spanners and keys*

Use of a piece of tubing to extend a spanner or key is very bad practice. It strains the jaws of the spanner so that it becomes loose and may slip. It may even crack the jaws of the spanner so that they break. In both cases this can lead to nasty injuries to your hands and even a serious fall if you are working on a ladder. Also it over stresses the fastening which will be weakened or even broken. Always check a spanner for damage and correct fit before using it. A torque spanner should be used to tighten important fastenings.

2.5.5 Drilling

Drilling is a process for producing holes. The holes may be cut from the solid or existing holes may be enlarged. The purpose of the drilling machine is to:

- Rotate the drill at a suitable speed for the material being cut and the diameter of the drill.
- Feed the drill into the workpiece.
- Support the workpiece being drilled; usually at right-angles to the axis of the drill. On some machines the table may be tilted to allow holes to be drilled at a pre-set angle.

Activity 2.12

Select suitable hand tools (giving reasons for your choice) and draw up a production plan for making the component shown in Figure 2.55. Present your work in the form of a brief report with hand drawn sketches.
Do not forget to include your report and sketches in your portfolio.

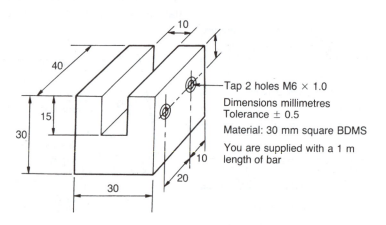

Figure 2.55 *See Activity 2.12*

Drilling machines

Drilling machines come in a variety of types and sizes. Figure 2.56 shows a hand held, electrically driven, power drill. It depends upon the skill of the operator to ensure that the drill cuts at right-angles to the workpiece. The feed force is also limited to the muscular strength of the user. Figure 2.57 shows a more powerful, floor mounted machine. The spindle rotates the drill. It can also move up and down in order to feed the drill into the workpiece and withdraw the drill at the end of

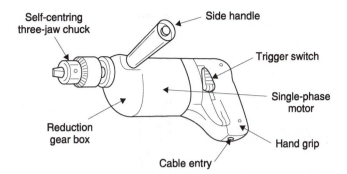

Figure 2.56 *An electric power drill*

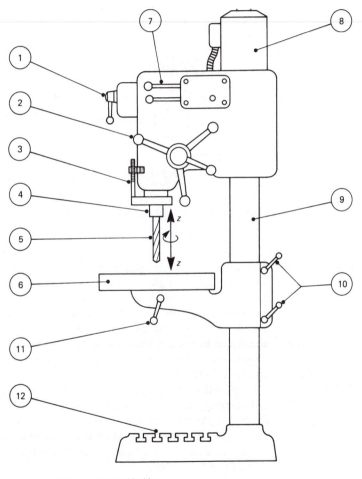

Parts of the Pillar Type Drilling Machine

1 Stop/start switch (electrics).
2 Hand or automatic feed lever.
3 Drill depth stop.
4 Spindle.
5 Drill.
6 Table.

7 Speed change levers.
8 Motor.
9 Pillar.
10 Vertical table lock.
11 Table lock.
12 Base.

Figure 2.57 *A pillar drill*

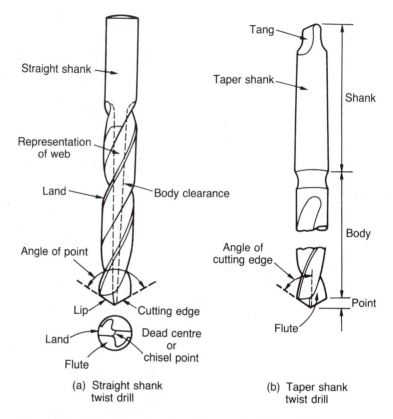

(a) Straight shank
    twist drill

(b) Taper shank
    twist drill

Figure 2.58 *Straight shank and taper shank twist drills*

the cut. Holes are generally produced with twist drills. Figure 2.58 shows a typical straight shank drill and a typical taper shank drill and names their more important features.

Large drills have taper shanks and are inserted directly into the spindle of the machine as shown in Figure 2.59(a). They are located and driven by a taper. The tang of the drill is for extraction purposes only. It does not drive the drill. The use of a *drift* to remove the drill is shown in Figure 2.59(b).

Small drills have straight (parallel) shanks and are usually held in a self-centring chuck. Such a chuck is shown in Figure 2.59(c). The chuck is tightened with the chuck key shown. SAFETY: The chuck key must be removed before starting the machine. The drill chuck has a taper shank which is located in, and driven by, the taper bore of the drilling machine spindle.

The cutting edge of a twist drill is wedge shaped, like all the tools we have considered so far. This is shown in Figure 2.60.

When regrinding a drill it is essential that the point angles are correct. The angles for general purpose drilling are shown in Figure 2.61(a). After grinding, the angles and lip lengths must be checked as shown in Figure 2.61(b). The point must be symmetrical. The effects of incorrect grinding are shown in Figure 2.61(c).

If the lip lengths are unequal, an oversize hole will be drilled when cutting from the solid. If the angles are unequal, then only one lip will

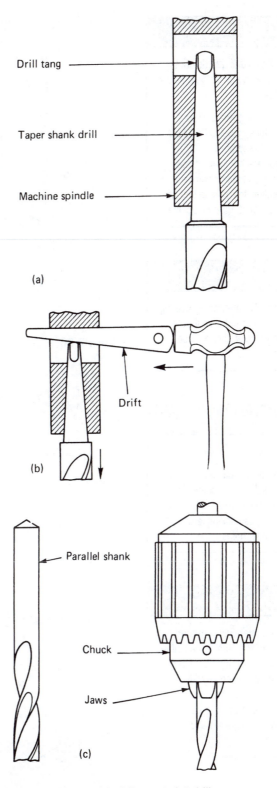

Drill tang

Taper shank drill

Machine spindle

(a)

Drift

(b)

Parallel shank

Chuck

Jaws

(c)

Figure 2.59 *Methods of holding a twist drill*

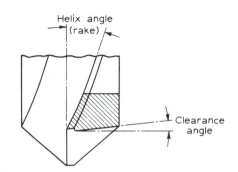

Figure 2.60 *Cutting edges of a twist drill*

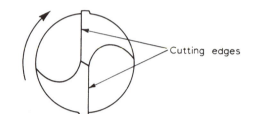

(a) Drill angles for general purpose drilling

(b) Checking for correct point angle and equal lip lengths

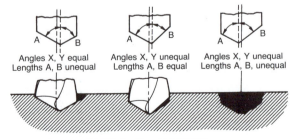

(c) Effects of incorrect grinding

Figure 2.61 *Point angles for a twist drill*

cut and undue wear will result. The unbalanced forces will cause the drill to flex and 'wander'. The axis of the hole will become displaced as drilling proceeds. If both these faults are present at the same time, both sets of faults will be present and an inaccurate and ragged hole will result.

Work holding
It is dangerous to hold work being drilled by hand. There is always a tendency for the drill to grab the work and spin it round. Also the rapidly spinning *swarf* can produce some nasty cuts to the back of your hand. Therefore, the work should always be securely fastened to the machine table.

Nevertheless, small holes in relatively large components are sometime drilled with the work handheld. In this case a stop bolted to the machine table should be used to prevent rotation.

Small work is usually held in a machine vice which, in turn, is securely bolted to the machine table as shown in Figure 2.62(a). Larger work can be clamped directly to the machine table as shown in Figure 2.62(b). In both these latter two examples the work is supported on parallel blocks. You mount the work in this way so that when the drill 'breaks through' the workpiece it does not damage the vice or the machine table.

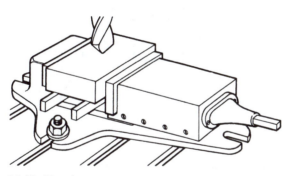

(a) Machine vice

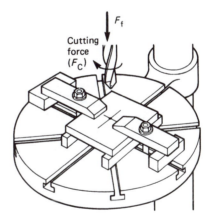

(b) Work supported on parallels and clamped to table

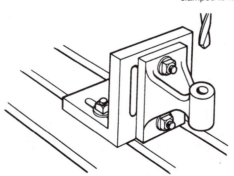

(c) Use of angle plate

Figure 2.62 *Work holding*

Figure 2.62(c) shows how an angle plate can be used when the hole axis has to be parallel to the datum surface of the work. Finally, Figure 2.63(a) and (b) show how cylindrical work is located and supported using vee-blocks.

Finally, Figure 2.64 shows some miscellaneous operations that are frequently carried out on drilling machines. These include countersinking, counterboring, and spot-facing.

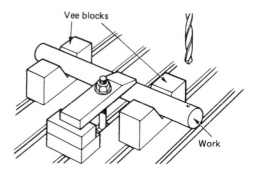

(a) HORIZONTAL

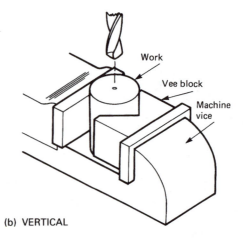

(b) VERTICAL

Figure 2.63 *Work holding cylindrical components*

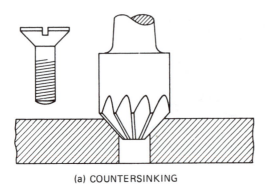

(a) COUNTERSINKING   (b) COUNTERBORING

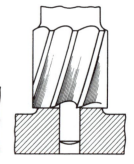

(c) SPOTFACING

Figure 2.64 *Countersinking, counterboring and spot facing*

Countersinking

Figure 2.64(a) shows a countersink bit being used to countersink a hole to receive the heads of rivets or screws. For this reason the included angle is 90°. Lathe centre drills are unsuitable for this operation as their angle is 60°.

Counterboring

Figure 2.64(b) shows a piloted counterbore being used to counterbore a hole so that the head of a capscrew or a cheese-head screw can lie below the surface of the work. Unlike a countersink cutter, a counterbore is not self-centring. It has to have a pilot which runs in the previously drilled bolt or screw hole. This keeps the counterbore cutting concentrically with the original hole.

Spot-facing

This is similar to counterboring but the cut is not as deep. It is used to provide a flat surface on a casting or a forging for a nut and washer to 'seat' on. Sometimes, as shown in Figure 2.64(c), it is used to machine a *boss* (raised seating) to provide a flat surface for a nut and washer to 'seat' on.

2.5.6 Centre lathe turning

The main purpose of a centre lathe is to produce external and internal cylindrical and conical (tapered) surfaces (Table 2.10). It can also produce plain surfaces and screw threads.

The centre lathe

Figure 2.65(a) shows a typical centre lathe and identifies its more important parts:

- The bed is the base of the machine to which all the other sub-assemblies are attached. Slideways accurately machined on its top surface provide guidance for the saddle and the tail stock.

Cutting movement	Hand or power traverse	Means by which movement is achieved	Turned feature
Tool parallel to the spindle centre line	Both	The saddle moves along the bed slideways	A parallel cylinder
Tool at 90° to the spindle centre line	Both	The cross-slide moves along a slideway machined on the top of the saddle	A flat face square to the spindle centre line
Tool at an angle relative to the spindle centre line	Hand	The compound slide is rotated and set at the desired angle relative to the centre line	A tapered cone

Table 2.10 *Centre lathe movements*

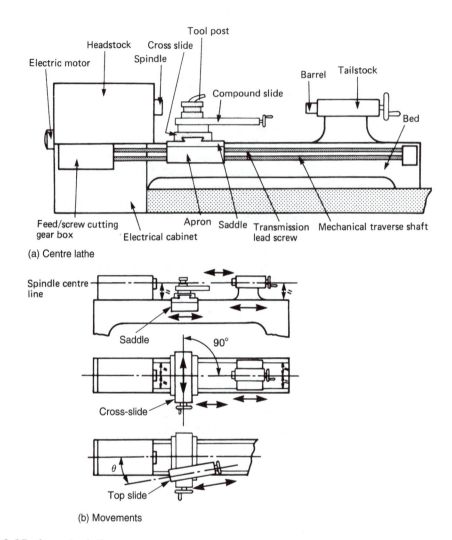

(a) Centre lathe

(b) Movements

Figure 2.65 *A centre lathe*

These slideways also locate the headstock so that the axis of the spindle is parallel with the movement of the saddle and the tailstock. The saddle or carriage of the lathe moves parallel to the spindle axis as shown in Figure 2.65(b).

- The cross-slide is mounted on the saddle of the lathe. It moves at 90° to the axis of the spindle as shown in Figure 2.65(c). It provides in-feed for the cutting tool when cylindrically turning. It is also used to produce a plain surface when facing across the end of a bar or component.

- The top-slide (compound-slide) is used to provide in-feed for the tool when facing. It can also be set at an angle to the spindle axis for turning tapers as shown in Figure 2.65(b).

Work holding in the lathe
The work to be turned can be held in various ways. We will now consider the more important of these.

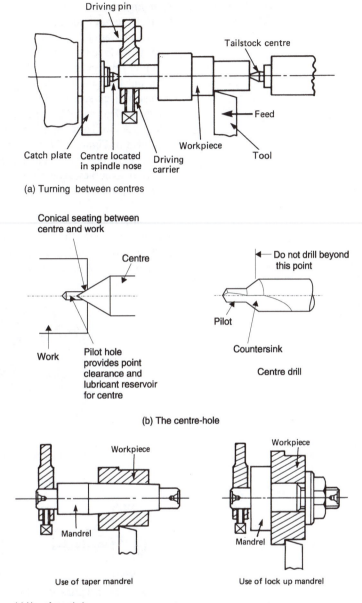

(a) Turning between centres

(b) The centre-hole

(c) Use of mandrels

Figure 2.66 *Work holding in a centre lathe*

Between centres

The centre lathe derives its name from this basic method of work holding. The general layout is shown in Figure 2.66(a). Centre holes are drilled in the ends of the bar and these locate on centres in the headstock spindle and the tailstock barrel. A section through a correctly centred component is shown in Figure 2.66(b). The centre-hole is cut with a standard centre-drill. The main disadvantage of this method of work holding is that no work can be performed on the end of the component. Work that has been previously bored can be finish turned between centres using a taper mandrel as shown in Figure 2.66(c).

Four-jaw chuck

Chucks are mounted directly onto the spindle nose and hold the work securely without the need for a back centre. This allows the end of the work to be faced flat. It also allows for the work to have holes bored into it or through it.

In the four-jaw chuck, the jaws can be moved independently by means of jack-screws. The jaws can also be reversed as shown in Figure 2.67(a) and the work held in various ways as shown in Figure 2.67(b). As well as cylindrical work, rectangular work can also be held as shown in Figure 2.67(c). As the jaws can be moved independently, the work can be set to run concentrically with the spindle axis to a high degree of accuracy. Alternatively, the work can be deliberately set off-centre to produce eccentric components as shown in Figure 2.67(d).

Three-jaw chuck

The self-centring, three-jaw chuck is shown in Figure 2.68(a). The jaws are set at 120° and are moved in or out simultaneously (at the same time) by a scroll when the key is turned. SAFETY: This key

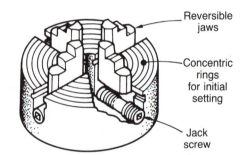

(a) Independent four-jaw chuck

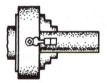

Jaws in normal position   Chuck jaws reversed   Work chucked on inside

(b) Methods of holding work in a chuck

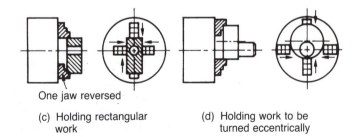

One jaw reversed

(c) Holding rectangular work

(d) Holding work to be turned eccentrically

Figure 2.67 *Four-jaw chuck*

must be removed before starting the lathe or a serious accident can occur. When new and in good condition this type of chuck can hold cylindrical and hexagonal work concentric with the spindle axis to a high degree of accuracy. In this case the jaws are not reversible, so it is provided with separate internal and external jaws. In Figure 2.68 (a) the internal jaws are shown in the chuck, and the external jaws are shown at the side of the chuck. Again the chuck is mounted directly on the spindle nose of the lathe.

Work to be turned between centres is usually held in a three-jaw chuck whilst the ends of the bar are faced flat and then centre drilled as shown in Figure 2.68(b).

Face-plate

Figure 2.69 shows a component held on a face-plate so that the hole can be bored perpendicularly to the datum surface. This datum surface is in contact with the face-plate. Note that the face-plate has to be balanced to ensure smooth running. Care must be taken to check that the clamps will hold the work securely and do not foul the machine. The clamps must not only resist the cutting forces, but they must also prevent the rapidly rotating work from spinning out of the lathe.

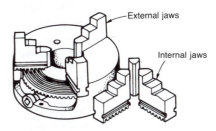

(a) Three jaw self-centring chuck

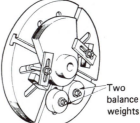

(b) Centring work held
in a three-jaw chuck

Figure 2.68 *Three-jaw chuck*

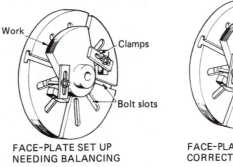

FACE-PLATE SET UP
NEEDING BALANCING

FACE-PLATE SET UP
CORRECTLY BALANCED

Figure 2.69 *Face-plate set up*

Turning tools

Figure 2.70(a) shows a range of turning tools and some typical applications. Figure 2.70(b) shows how the metal-cutting wedge also applies to turning tools. Turning tools are fastened into a tool-post which is mounted on the top-slide of the lathe. There are many different types of tool-post. The four-way turret tool-post shown in Figure 2.70(c) allows four tools to be mounted at any one time.

Parallel turning

Figure 2.71(a) shows a long bar held between centres. To ensure that the work is truly cylindrical with no taper, the axis of the tailstock centre must be in line with the axis of the headstock spindle. The saddle traverse provides movement of the tool parallel with the workpiece axis. You take a test cut and measure the diameter of the bar at both ends. If all is well, the diameter should be constant all along the bar. If not, the lateral movement of the tailstock needs to be adjusted until a constant measurement is obtained. The depth of cut is controlled by micrometer adjustment of the cross-slide.

Whilst facing and centre drilling the end of a long bar, a fixed steady is used. This supports the end of the bar remote from the chuck. A fixed steady is shown in Figure 2.71(b).

If the work is long and slender it sometimes tries to kick away from the turning tool or even climb over the tool. To prevent this happening a travelling steady is used. This is bolted to the saddle opposite to the tool as shown in Figure 2.71(c).

Surfacing

A surfacing (facing or perpendicular turning) operation on a workpiece held in a chuck is shown in Figure 2.72. The saddle is clamped to the bed of the lathe and the tool motion is controlled by the cross-slide. This ensures that the tool moves in a path at right-angles to the workpiece axis and produces a plain surface. In-feed of the cutting tool is controlled by micrometer adjustment of the top-slide.

Boring

Figure 2.73 shows how a drilled hole can be opened up using a boring tool. The workpiece is held in a chuck and the tool movement is controlled by the saddle of the lathe. The in-feed of the tool is controlled by micrometer adjustment of the cross-slide. The pilot hole is produced either by a taper shank drill mounted directly into the tailstock barrel (poppet), or by a parallel shank drill held in a drill chuck. The taper mandrel of the drill chuck is inserted into the tailstock barrel.

Conical surfaces

Chamfers on the corners of a turned component are short conical surfaces. These are usually produced by using a chamfering tool as shown in Figure 2.70(a). Longer tapers can be produced by use of the top-slide. Use of the top (compound) slide is shown in Figure 2.74. The slide is mounted on a swivel base and it is fitted with a protractor scale. It can be swung round to the required angle and clamped in position. The taper is then cut as shown.

Miscellaneous turning operations

Reamers are sizing tools. They remove very little metal. Since they follow the existing hole, they cannot correct the positional errors.

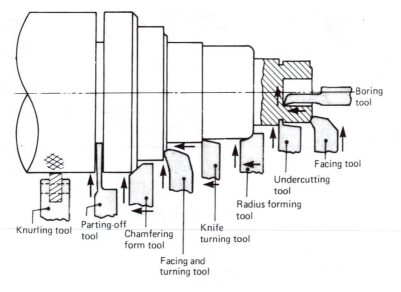

(a) Turning tools

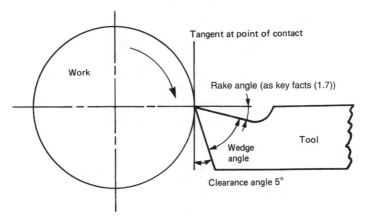

(b) Turning tool angles

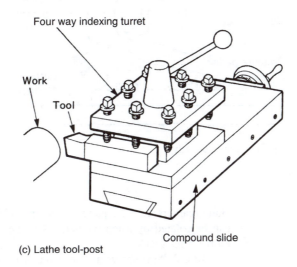

(c) Lathe tool-post

Figure 2.70 *Turning tools*

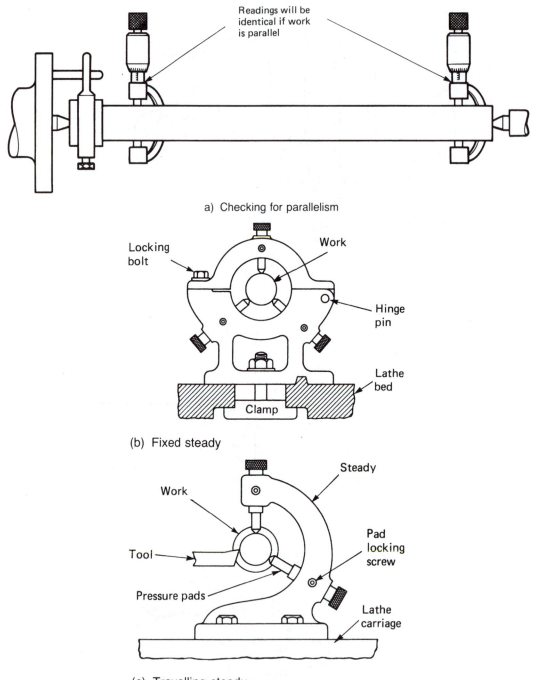

a) Checking for parallelism

(b) Fixed steady

(c) Travelling steady

Figure 2.71 *Parallel turning*

Hand reamers have a square on the end of their shanks so that they can be rotated by a tap wrench. Machine reamers have a standard morse taper shank.

Figure 2.75(a) shows a hole being reamed in a lathe. A machine reamer is being used and it is held in the barrel of the tailstock. As a

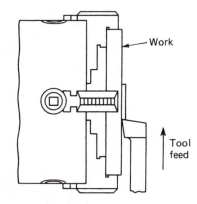

Figure 2.72 *Surfacing*

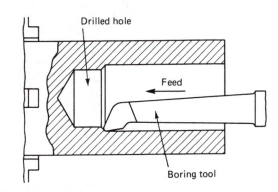

Figure 2.73 *Boring*

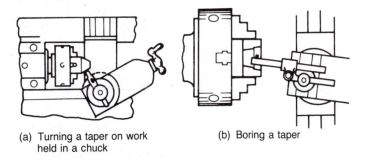

(a) Turning a taper on work held in a chuck

(b) Boring a taper

Figure 2.74 *Producing a chamfer*

drilled hole invariably runs out slightly, the pilot hole should be single-point bored in order to correct the position and geometry of the hole. It is finally sized using the reamer. Only the minimum amount of metal for the surface of the hole to clean up should be left in for the reamer to remove.

Standard, non-standard and large diameter screw threads can be cut in a centre lathe by use of the lead-screw to control the saddle movement. This is a highly skilled operation. However, standard screw threads of limited diameter can be cut using hand threading tools as shown in Figure 2.75(b) and (c). Taps are very fragile and the

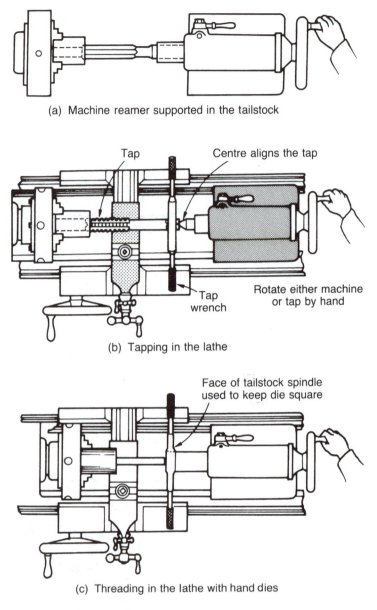

(a) Machine reamer supported in the tailstock

(b) Tapping in the lathe

(c) Threading in the lathe with hand dies

Figure 2.75 *Miscellaneous turning operations*

workpiece should be rotated by hand with the lathe switched off and the gears disengaged.

Activity 2.13

Select suitable hand tools (giving reasons for your choice) and draw up a production plan for making the component shown in Figure 2.76. Present your work in the form of a brief report with hand drawn sketches. Do not forget to include your report and sketches in your portfolio.

Test your knowledge 2.54

List FIVE main operations that can be performed on a centre lathe.

Test your knowledge 2.55

Describe, with the aid of sketches, THREE different types of turning tool for use with a centre lathe.

Test your knowledge 2.56

Describe, with the aid of sketches, TWO different types of chuck for use with a centre lathe.

Test your knowledge 2.57

With the aid of a sketch explain how you would hold the component shown in Figure 2.76(b) in order to machine the 50 mm diameter hole.

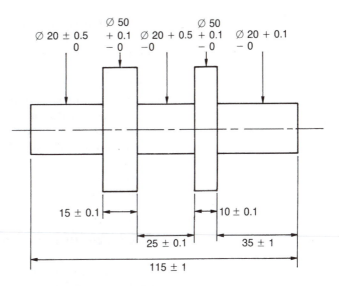

Dimensions in millimetres
Material: Phosphor bronze: Blank size Ø 60 × 125
Hint: Turn between centres, after facing and centring

(a)

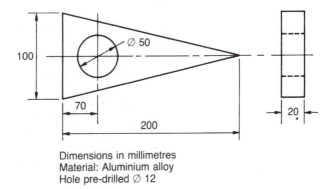

Dimensions in millimetres
Material: Aluminium alloy
Hole pre-drilled Ø 12

(b)

Figure 2.76 *See Activity 2.13 and Test your knowledge 2.57*

2.5.7 Casting and milling

Casting is the name given to the process of shaping metal by pouring hot metal into a mould. Once the material has cooled and solidified is can be removed from the mould whereupon it will have a shape defined by the cavity in which it has solidified.

Different methods of casting are in common use. In *sand casting* the mould is formed by sand placed in a pair of mating boxes in which a wooden model of the component has been used to define the shape of the mould in the sand. In *die-casting* the mould shape is formed in metal. Castings can be made from steel, cast iron, brass, aluminium as well as most other metals and alloys.

Cast components often require further finishing before they can be assembled into finished products. One reason for this is that casting is not a particularly precise process and the dimensions of cast components can often vary slightly. Another reason is that the surface of a cast component may not be perfectly flat. This is a particularly important consideration where parts need to fit together.

Cast components (or at least some surfaces of cast components) are often *milled* in order to provide a more precise surface and to remove any irregularities. Milling is a process carried out by a *milling machine* which uses a rotating toothed cutting wheel. Milling machines are available in two forms: *horizontal mills* and *vertical mills* according to the direction of the cutting axis.

2.5.8 Joining and assembly

Engineered products usually comprise a number of components that must be assembled or joined together in some particular way. The purpose of assembly is to put together a number of individual components to build up a whole device, structure or system. To achieve this aim attention must be paid to the following key factors.

Sequence of assembly
This must be planned so that as each component is added, its position in the assembly and the position of its fastenings are accessible. Also the sequence of assembly must be planned so that the installation of one component does not prevent access for fitting the next component or some later component.

Technique of joining
These must be selected to suit the components being joined, the materials from which they are made, and what they do in service. If the joining technique involves heating, then care must be taken that adjacent components are not heat sensitive or flammable.

Position of joints
Joints must not only be accessible for initial assembly, they must also be accessible for maintenance. You do not want to dismantle half a machine to make a small adjustment, or replace a part that wears out regularly.

Interrelationship and identification of parts
Identification of parts and their position in an assembly can usually be determined from assembly drawings or exploded view drawings. Interrelationship markings are often included on components. For example, the various members and joints of structural steelwork are given number and letter codes to help identification on site. PCBs usually have the outline of the various components printed on them as well as the part number.

Tolerances
The assembly technique must take into account the accuracy and finish of the components being assembled. Much greater care has to be

taken when assembling a precision machine tool or an artificial satellite, than when assembling structural steel work.

Protection of parts

Components awaiting assembly require protection against accidental damage and corrosion. In the case of structural steelwork this may merely consist of painting with red oxide primer and careful stacking. Precision components will require treating with an anti-corrosion lanolin based compound that can be easily removed at the time of assembly. Bores must be sealed with plastic plugs and screw threads with plastic caps. Precision ground surfaces must also be protected from damage. Heavy components must be provided with eye-bolts for lifting. Vulnerable sub-assemblies such as aircraft engines must be supported in suitable cradles.

Joining (mechanical)

The joints used in engineering assemblies may be divided into the following categories.

Permanent joints

These are joints in which one or more of the components and/or the joining medium has to be destroyed or damaged in order to dismantle the assembly; for example, a riveted joint.

Temporary joints

These are joints that can be dismantled without damage to the components. It should be possible to re-assemble the components using the original or new fastenings; for example, a bolted joint.

Flexible joints

These are joints in which one component can be moved relative to another component in an assembly in a controlled manner; for example, the use of a hinge.

These are used to make temporary joints that can be dismantled and re-assembled at will. They are required where maintenance is necessary. We considered different types of screwed fastenings in the section on component selection. We also considered the different types of head found on screwed fastenings. Figure 2.77 shows the correct way to use some typical screwed fastenings.

Screwed fastenings must always pull down onto prepared seatings that are flat and at right-angles to the axis of the fastening. This prevents the bolt or screw being bent as it is tightened up. To protect the seating, a soft washer is placed between the seating and the nut. Taper washers are used when erecting steel girders to prevent the draught angle of the flanges from bending the bolt.

Locking devices are used to prevent screwed fastenings from slackening off due to vibration. Locking devices may be frictional or positive. A selection of plain washers, taper washers and locking devices is shown in Figure 2.78.

Riveting

The selection of rivets and rivet heads was considered in the section on component selection. However, to make a satisfactory riveted joint the following points must be observed.

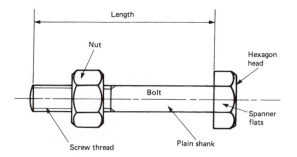

(a) HEXAGON HEAD BOLT AND NUT

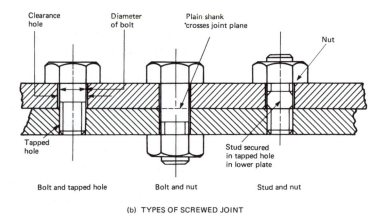

(b) TYPES OF SCREWED JOINT

Figure 2.77 *Screwed fastenings*

Hole clearance

If the clearance is too small there will be difficulty in inserting the rivet and drawing up the joint. If the hole is too large, the rivet will buckle and a weak joint will result.

Rivet length

If the rivet is too long, the rivet bends over during heading. If the rivet is too short the head cannot be properly formed. In either case a weak joint will result. Figure 2.79(a) shows the correct proportions for a riveted joint and Figure 2.79(b) shows some typical riveting faults. The correct procedure for heading (closing) a rivet is shown in Figure 2.80(a). The drawing up tool ensures that the components to be drawn are brought into close contact and that the head of the rivet is drawn up tightly against the lower component.

The hammer blows with the flat face of the hammer swells the rivet so that it is a tight fit in the hole and starts to form the head. The ball pein of the hammer head is then used to rough form the rivet head. The head is finally finished and made smooth by using a rivet snap. Where large rivets and large quantities of rivets are to be closed a portable pneumatic tool is used.

Fibre or plastic insert
(frictional locking)

Slot for split pin
(positive locking)

SELF-LOCKING NUT

CASTLE NUT

PLAIN WASHER

SPRING WASHER
(FRICTIONAL LOCKING)

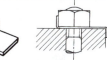

TAB WASHER

TAB WASHER IN USE
(POSITIVE LOCKING)

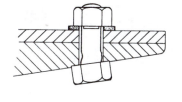

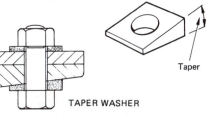

Taper

TAPER WASHER

5° TAPER FOR USE ON CHANNEL ⎫ FLANGES
8° TAPER FOR USE ON JOIST ⎭

Figure 2.78 *Various nuts and washers*

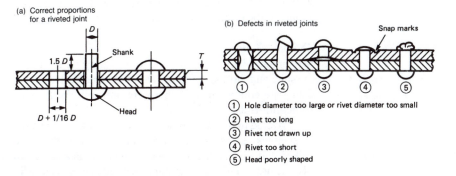

(a) Correct proportions
for a riveted joint

Shank

1.5 D

D

T

Head

$D + 1/16\ D$

(b) Defects in riveted joints

Snap marks

① ② ③ ④ ⑤

① Hole diameter too large or rivet diameter too small
② Rivet too long
③ Rivet not drawn up
④ Rivet too short
⑤ Head poorly shaped

Figure 2.79 *Correct and incorrect riveted joints*

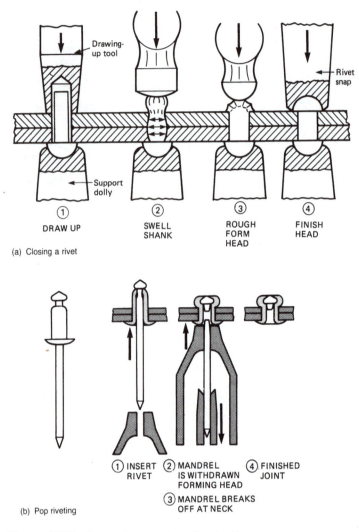

Figure 2.80 *Correct procedure for riveting*

'Pop' riveting is often used for joining thin sheet metal components, particularly when building up box sections. When building up box sections it is not possible to get inside the closed box to use a hold-up dolly; for example, when riveting the skin to an aircraft wing. Figure 2.80(b) shows the principle of 'pop' riveting.

Fusion welding
Welding has largely taken over from riveting for many purposes such as ship and bridge building and for structural steelwork. Welded joints are continuous and, therefore, transmit the stresses across the joint uniformly.

In riveted joints the stresses are concentrated at each rivet. Also the rivet holes reduce the cross-sectional areas of the members being joined and weaken them. However, welding is a more skilled assembly technique and the equipment required is more costly. The components being joined are melted at their edges and additional filler metal

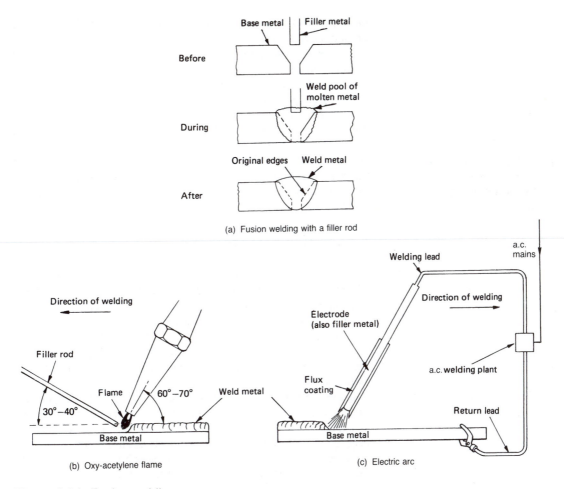

(a) Fusion welding with a filler rod

(b) Oxy-acetylene flame

(c) Electric arc

Figure 2.81 *Fusion welding*

is melted into the joint. The filler metal is of similar composition to that of the components being joined. Figure 2.81(a) shows the principle of fusion welding.

High temperatures are involved to melt the metal of the components being joined. These can be achieved by using the flame of an oxy-acetylene blowpipe as shown in Figure 2.81(b), or an electric arc as shown in Figure 2.81(c). When oxy-acetylene welding (gas welding), a separate filler rod is used. When arc welding, the electrode is also the filler rod and is melted as welding proceeds.

No flux is required when oxy-acetylene welding as the molten metal is protected from atmospheric oxygen by the burnt gases (products of combustion). When arc welding, a flux is required. This is in the form of a coating surrounding the electrode. This flux coating is not only deposited on the weld to protect it, it also stabilises the arc and makes the process easier. The hot flux gives off fumes and adequate ventilation is required.

Protective clothing must be worn when welding and goggles or a face mask (visor) appropriate for the process must be used. These have optical filters that protect the user's eyes from the harmful radiations produced during welding. The optical filters must match the process.

The compressed gases used in welding are very dangerous and welding equipment must only be used by skilled persons or under close supervision. Acetylene gas bottles must only be stored and used in an upright position.

The heated area of the weld is called the weld zone. Because of the high temperatures involved, the heat affected area can spread back into the parent metal of the component for some distance from the actual weld zone. This can alter the structure and properties of the material so as to weaken it and make it more brittle. If the joint fails in service, failure usually occurs at the side of the weld in this heat affected zone. The joint itself rarely fails.

Soft soldering

Soft soldering is also a thermal jointing process. Unlike fusion welding, the parent metal is not melted and the filler metal is an alloy of tin and lead that melts at relatively low temperatures. Soft soldering is mainly used for making mechanical joints in copper and brass components (plumbing). It is also used to make permanent electrical connections. Low carbon steels can also be soldered providing the metal is first cleaned and then *tinned* using a suitable flux. The tin in the solder reacts chemically with the surface of the component to form a bond.

Figure 2.82 shows how to make a soft soldered joint. The surfaces to be joined are first degreased and physically cleaned to remove any dust and dirt. Fine abrasive cloth or steel wool can be used. A flux is used to render the joint surfaces chemically clean and to make the solder spread evenly through the joint. Some soft soldering fluxes and their typical applications are listed in Table 2.11.

- The copper *bit* of the soldering iron is then heated. For small components and fine electrical work an electrically heated iron can be used. For joints requiring a soldering iron with a larger bit, a gas heated soldering stove can be used to heat the bit.

- The heated bit is then cleaned, fluxed and coated with solder. This is called *tinning* the bit.

- The heated and tinned bit is drawn slowly along the fluxed surfaces of the components to be joined. This transfers solder to the surfaces of the components. Additional solder can be added if required. The work should be supported on wood to prevent heat loss. The solder does not just 'stick' to the surface of the metal being tinned. The solder reacts chemically with the surface to form an amalgam that penetrates into the surface of the metal. This forms a permanent bond.

- Finally the surfaces are overlapped and 'sweated' together. That is, the soldering iron is re-heated and drawn along the joint as shown. Downward pressure is applied at the same time. The solder in the joint melts. When it solidifies it forms a bond between the two components.

Test your knowledge 2.58

State THREE advantages and THREE disadvantages of welding compared with riveting.

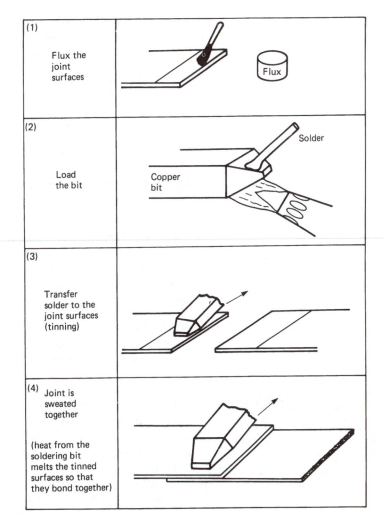

Figure 2.82 *Procedure for making a soft soldered joint*

Flux	Metals	Characteristics
Killed spirits (acidulated zinc chloride solution)	Steel, tin plate, brass and copper	Powerful cleansing action but leaves corrosive residue
Dilute hydrochloric acid	Zinc and galvanised iron	As above, wash after use
Resin paste or 'cored' solder	Electrical conductor and terminal materials	Only moderate cleansing action (passive flux) but non-corrosive
Tallow	Lead and pewter	As above
Olive oil	Tin plate	Non-toxic, passive flux for food containers, non-corrosive

Table 2.11 *Soldering fluxes*

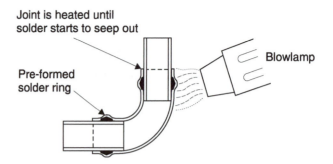

Figure 2.83 *'Sweating' a copper pipe*

Figure 2.83 shows how a copper pipe is sweated to a fitting. The pipe and the fitting are cleaned, fluxed and assembled. The joint is heated with a propane gas torch and solder is added. This is usually a resin-flux cored solder. The solder is drawn into the close fitting joint by capillary action.

Hard soldering
Hard soldering uses a solder whose main alloying elements are copper and silver. Hard soldering alloys have a much higher melting temperature range than soft solders. The melting range for a typical soft solder is 183–212°C. The melting range for a typical hard solder is 620–680°C. Hard soldering produces joints that are stronger and more ductile. The melting range for hard solders is very much lower than the melting point of copper and steel, but it is only just below the melting point of brass. Therefore, great care is required when hard soldering brass to copper. Because the hard solder contains silver it is often referred to as 'silver solder'. A special flux is required based on borax.

A soldering iron cannot be used because of the high temperatures involved. Heating is by a blow pipe. Figure 2.84 shows you how to make a typical hard soldered joint. Again cleanliness and careful

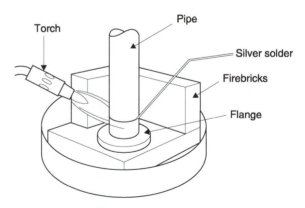

- The work is up to heat when the silver solder melts on contact with the work with the flame momentarily withdrawn.
- Add solder as required until joint is complete.

Figure 2.84 *Procedure for making a hard soldered joint*

> **Test your knowledge 2.59**
>
> List the advantages and limitations of soft soldering compared with hard soldering.

surface preparation are essential for a successful joint. The joint must be close fitting and free from voids. The silver solder is drawn into the joint by capillary action.

Even stronger joints can be made using a brass alloy instead of a silver–copper alloy. This is called *brazing*. The temperatures involved are higher than those for silver soldering. Therefore, brass cannot be brazed. The process of brazing is widely used for joining the steel tubes and malleable cast iron fittings of bicycle frames.

Activity 2.14

Consult manufacturers' or suppliers' data and draw up a table showing the composition of several soft solders and their typical applications. Present your work in the form of an overhead projector transparency and include this in your portfolio.

Adhesive bonding
The advantages of adhesive bonding can be summarised as follows.

- The temperature rise from the curing of the adhesive is negligible compared with that of welding. Therefore, the properties of the materials being joined are unaffected.
- Similar and dissimilar materials can be joined.
- Adhesives are electrical insulators. Therefore, they reduce or prevent electrolytic corrosion when dissimilar metals are joined together.
- Joints are sealed and fluid tight.
- Stresses are transmitted across the joint uniformly.
- Depending upon the type of adhesive used, some bonded joints tend to damp out vibrations.

Bonded joints have to be specially designed to exploit the properties of the adhesive being used. You cannot just substitute an adhesive in a joint designed for welding, brazing or soldering. Figure 2.85(a) shows some typical bonded joint designs that provide a large contact area. A correctly designed bonded joint is very strong. Major structural members in modern high performance airliners and military aircraft are adhesive bonded. Figure 2.85(b) defines some of the jargon used when talking about bonded joints. The strength of a bonded joint depends upon two factors, adhesion and cohesion.

Adhesion
This is the ability of the adhesive to 'stick' to the materials being joined (the *adherends*). This can result from physical keying or interlocking as shown in Figure 2.86(a). Alternatively specific bonding can take place. Here, the adhesive reacts chemically with the surface of the adherends as shown in Figure 2.86(b). Bonding occurs through intermolecular attraction.

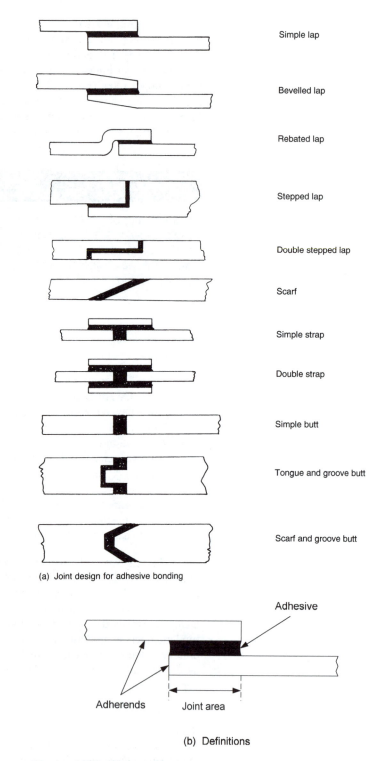

Simple lap

Bevelled lap

Rebated lap

Stepped lap

Double stepped lap

Scarf

Simple strap

Double strap

Simple butt

Tongue and groove butt

Scarf and groove butt

(a) Joint design for adhesive bonding

Adhesive

Adherends

Joint area

(b) Definitions

Figure 2.85 *Typical bonded joints*

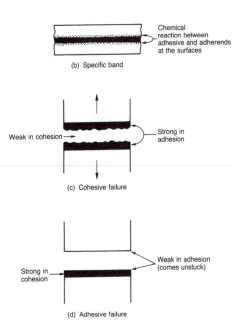

(a) Mechanical interlocking in porous materials

(b) Specific band

Chemical reaction between adhesive and adherends at the surfaces

Weak in cohesion

Strong in adhesion

(c) Cohesive failure

Weak in adhesion (comes unstuck)

Strong in cohesion

(d) Adhesive failure

Figure 2.86 *Adhesion*

Cohesion

This is the internal strength of the adhesive. It is the ability of the adhesive to withstand forces within itself. Figure 2.86(c) shows the failure of a joint made from an adhesive that is strong in adhesion but weak in cohesion. Figure 2.86(d) shows the failure of a joint that is strong in cohesion but weak in adhesion.

As well as the design of the joint, the following factors affect the strength of a bonded joint:

- The joint must be physically clean and free from dust, dirt, moisture, oil and grease.
- The joint must be chemically clean. The materials being joined must be free from scale or oxide films.
- The environment in which bonding takes place must have the correct humidity and be at the correct temperature.

Bonded joints may fail in four ways. These are shown in Figure 2.87. Bonded joints are least likely to fail in tension and shear. They are most likely to fail in cleavage and peel.

The most efficient way to apply adhesives is by an adhesive gun. This enables the correct amount of adhesive to be applied to the correct place without wastage or mess. It also prevents the evaporation of highly flammable and toxic solvents whilst the adhesive is waiting to be used.

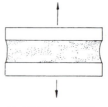

(a) JOINT IN TENSION

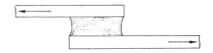

(b) JOINT IN SHEAR

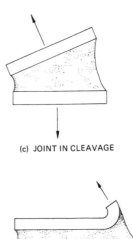

(c) JOINT IN CLEAVAGE

(d) JOINT IN PEEL

Figure 2.87 *Ways in which bonded joints can fail*

Joining (electrical and mechanical)

Again, joints may be permanent or temporary. Permanent joints are soldered or crimped. Temporary joints are bolted, clamped or plugged in.

Soldered joints

When soldering electrical and electronic components it is important not to overheat them. Overheating can soften thermoplastic insulation and completely destroy solid-state devices such as diodes and transistors. Very often some form of heatsink is required when soldering solid-state devices.

A high tin content low melting temperature solder with a resin flux core should be used. This is a passive flux. It only protects the joint. It contains no active, corrosive chemicals to clean the joint. Therefore, the joint must be kept clean whilst soldering. Even the natural grease from your fingers is sufficient to cause a high resistance 'dry' joint.

Test your knowledge 2.60

Describe the essential differences between soldering and brazing.

Test your knowledge 2.61

List the safety precautions that must be taken when using oxyacetylene equipment.

Test your knowledge 2.62

With the aid of sketches describe THREE faults that can occur when making riveted joints.

Test your knowledge 2.63

A bolt is described as 'M5 × 0.8'. Explain what this means.

Test your knowledge 2.64

State the precautions during the design stage and during the manufacture of an adhesive bonded joint.

Test your knowledge 2.65

Describe the advantages of an active soldering flux compared with a passive soldering flux.

Figure 2.88(a) shows how a soldered connection is made to a solder tag. Note how the lead from the resistor is secured around the tag before soldering. This gives mechanical strength to the connection. Soldering provides the electrical continuity.

Figure 2.88(b) shows a prototype electronic circuit assembled on a matrix board. The board is made from laminated plastic and is pierced with a matrix of equally spaced holes. Pin tags are fastened into these holes in convenient places and the components are soldered to these pin tags.

Figure 2.88(c) shows the same circuit built up on a strip board. This is a laminated plastic board with copper tracks on the underside. The wire tails from the components pass through the holes in the board and are soldered to the tracks on the underside. The copper tracks are cut wherever a break in the circuit is required.

Figure 2.88(d) shows the underside of a PCB. This is built up as shown in Figure 2.88(c), except that the tracks do not need to be cut since they are customised for the circuit.

Large volume assembly of PCBs involves the use of pick and place robots to install the components. The assembled boards are then carried over a flow soldering tank on a conveyor. A roller rotates in the molten solder creating a 'hump' in the surface of the solder. As the assembled and fluxed board passes over this 'hump' of molten solder the components tags are soldered into place.

Wire wrapping

Wire wrapping is widely used in telecommunications where large numbers of fine conductors have to be terminated quickly and in close proximity to each other. Soldering would be inconvenient and the heat could damage the insulation of adjoining conductors. Also soldered joints would be difficult to disconnect. A special tool is used that automatically strips the insulation from the wire and binds the wire tightly around the terminal pins. The terminal pins are square in section with sharp corners. The corners cut into the conductor and prevent it from unwinding. The number of turns round the terminal is specified by the supervising engineer.

Activity 2.15

A manufacturer of electronic kits has asked you to produce a single page instruction sheet on 'How to solder your kit'. Produce a word processed instruction sheet and illustrate it using appropriate sketches and drawings. Include a section headed 'Safety'.

Crimped joints

For power circuits, particularly in the automotive industry, cable lugs and plugs are crimped onto the cables. The sleeve of the lug or the plug is slipped over the cable and then indented by a small pneumatic or hydraulic press. This is quicker than soldering and, as no heat is involved, there is no danger of damaging the insulation. Portable equipment is also available for making crimped joints on site. Hand

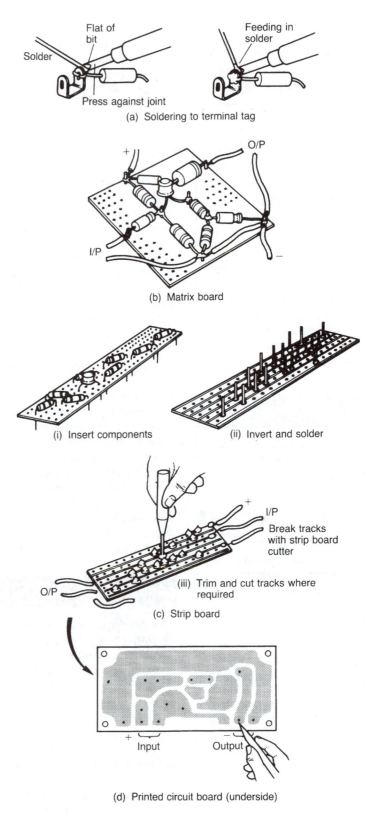

(a) Soldering to terminal tag

(b) Matrix board

(i) Insert components

(ii) Invert and solder

(iii) Trim and cut tracks where required

(c) Strip board

(d) Printed circuit board (underside)

Figure 2.88 *Various methods of electronic circuit assembly*

operated equipment can be used to fasten lugs to small cables by crimping as shown in Figure 2.89.

Clamped connections

Finally, we come to clamped connections using screwed fastenings. You will have seen many of these in domestic plugs, switches and lamp-holders. For heavier power installations, cable lugs are bolted to solid copper bus-bars using brass or bronze bolts as shown in Figure 2.90.

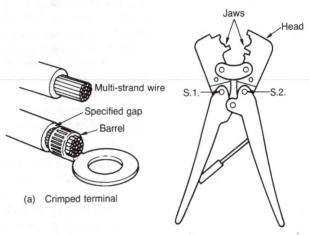

Figure 2.89 *Crimping*

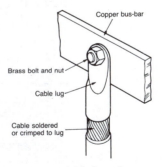

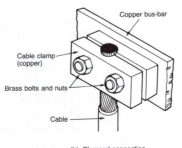

Figure 2.90 *Bolted and clamped connections*

2.5.9 Heat treatment

The properties of many metals and alloys can be changed by heating them to specified temperatures and cooling them under controlled conditions at specified rates. These are called, respectively, critical temperatures and critical cooling rates. We are only going to consider the heat treatment of plain carbon steels.

Hardening
The degree of hardness that can be given to any plain carbon steel depends upon two factors: the amount of carbon present, and how quickly the steel is cooled from the hardening temperature. The hardening temperature for medium carbon steels containing up to 0.8% carbon is bright red heat. Above 0.8% carbon the hardening temperature is dull red (cherry red) heat. Table 2.12(a) relates hardness to carbon content. Table 2.12(b) relates hardness to rate of cooling.

If oil quenching is used, a number of safety rules must be observed because of its flammability.

- Never use lubricating oil
- Always use a good quality quenching oil
- Always use a metal quenching bath with an airtight metal lid to smother the flames should the oil ignite
- Always keep the bath covered when not in use to keep out foreign objects and to avoid the absorption of moisture from the atmosphere.

Under heating
If the temperature of a steel does not reach its critical temperature, the steel will not harden.

Type of steel	Carbon content (%)	Effect of quench hardening
Low carbon	<0.3	Negligible
Medium carbon	0.3–0.5	Becomes tougher
S	0.5–0.9	Becomes hard
High carbon	0.9–1.3	Becomes very hard

(a) Effect of carbon content

Carbon content (%)	Quenching media	Required treatment
0.3–0.5	Oil	Toughening
0.5–0.9	Oil	Toughening
0.5–0.9	Water	Hardening
0.9–1.2	Oil	Hardening

(b) Rate of cooling

Table 2.12 *Heat treatment of steel*

Overheating

It is a common mistake to think that increasing the temperature from which the steel is quenched will increase its hardness. Once the correct temperature has been reached, the hardness will depend only upon the carbon content of the steel and its rate of cooling. If the temperature of a steel exceeds its critical temperature, grain growth will occur and the steel will be weakened. Also overheating will slow the cooling rate and will actually reduce the hardness of the steel.

Cracking

There are many causes of hardening cracks. Some of the more important are sharp corners, sudden changes of section, screw threads, holes too near the edge of a component. These should all be avoided at the design stage as should over rapid cooling for the type of steel being used.

Distortion

There are many causes of distortion. Some of the more important are as follows:

- Lack of balance or symmetry in the shape of the component.
- Lack of uniform cooling. Long thin components should be dipped end-on into the quenching bath.
- Change in the grain structure of the steel causing shrinkage.

No matter how much care is taken when quench hardening, some distortion (movement) will occur. Also slight changes in the chemical composition may occur at the surface of the metal. Therefore, precision components should be finish ground after hardening. The components must be left slightly oversize before grinding to allow for this. That is, a grinding allowance must be left on such components before hardening.

Tempering

Quench-hardened plain carbon steels are very brittle and unsuitable for immediate use. Therefore, further heat treatment is required. This is called tempering. It greatly increases the toughness of the hardened steel at the expense of some loss of hardness.

Tempering consists of re-heating the hardened steel and again quenching it in oil or water. Typical tempering temperatures for various applications are summarised in Table 2.13.

You can judge the tempering temperature by the colour of the oxide film. First, the component must be polished after hardening and before tempering. Then heat the component gently and watch for the colour of the metal surface to change. When you see the appropriate colour appear, the component must be quenched immediately.

2.5.10 Chemical treatment

The chemical treatments that will be considered in this section are as follows:

- Chemical machining (etching) as used in the production of PCBs
- The coating of metal components with decorative and/or corrosion resistant finishes.

Component	Temper colour	Temperature (°C)
Edge tools	Pale straw	220
Turning tools	Medium straw	230
Twist drills	Dark straw	240
Taps	Brown	250
Press tools	Brown/purple	260
Cold chisels	Purple	280
Springs	Blue	300
Toughening (medium carbon steels)	–	450–600

Table 2.13 *Tempering temperatures*

Test your knowledge 2.66

Describe how you should harden and temper a small cold chisel made from 1.2% high carbon steel.

Etching

PCBs have already been introduced in the section on assembly. First the circuit is drawn out by hand or designed using a computer aided design (CAD) package. A typical circuit master drawing is shown in Figure 2.91. The master drawing of the circuit is then photographed to produce a transparent copy called a negative. In a negative copy the light and dark areas are reversed. The PCB is made as follows:

• The copper coated laminated plastic (Tufnol) or fibre glass board is coated with a *photoresist* by dipping or spraying.

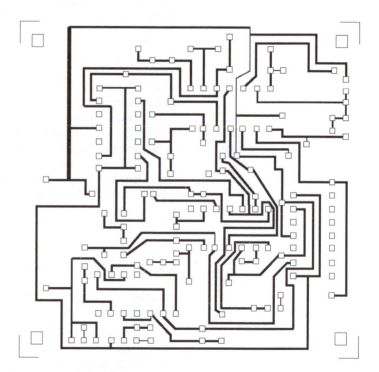

Figure 2.91 *A PCB layout*

- The negative of the circuit is placed in contact with the prepared circuit board. They are then exposed to ultraviolet light. The light passes through the transparent parts of the negative. The areas of the board exposed to the ultraviolet light will eventually become the circuit.

- The exposed circuit board is then developed in a chemical solution that hardens the exposed areas.

- The photoresist is stripped away from the unexposed areas of the circuit board.

- The circuit board is then placed in a suitable *etchant*. Ferric chloride solution can be used as an etchant for copper. This eats away the copper where it is not protected by the hardened photoresist. The remaining copper is the required circuit

- The circuit boards are washed to stop the reaction. The remaining photoresist is removed so as not to interfere with the tinning of the circuit with soft solder and the soldering of the components into position.

SAFETY: This process is potentially dangerous. Ultraviolet light is very harmful to your skin and to your eyes. The ferric chloride solution is highly corrosive to your skin. The various processes also give off harmful fumes. Therefore, you should only carry out this process under properly supervised, controlled and ventilated conditions. Appropriate protective clothing should be worn.

A similar process can be used for the chemical engraving of components with their identification numbers and other data.

Electroplating

The component to be plated is placed into a plating bath as shown in Figure 2.92. The component is connected to the negative terminal of a direct current supply. This operates at a low voltage but relatively heavy current. The anode is connected to the positive terminal of the supply. The anode is usually made from the same metal as that which is to be plated onto the component. The electrolyte is a solution of chemical salts. The composition depends upon the process being carried out.

When the current passes through the bath the component is coated with a thin layer of the protective and/or decorative metal. This metal comes from the chemicals in the electrolyte. The process is self-balancing. The anode dissolves into the electrolyte at the same rate as the metal taken from the electrolyte is being deposited on the component. This applies to most plating processes such as zinc, copper, tin and nickel plating.

An exception is chromium plating. A neutral anode is used that does not dissolve into the electrolyte. Additional salts have to be added to the electrolyte from time to time to maintain the solution strength. Chromium is not usually deposited directly onto the component. The component is usually nickel plated and polished and then a light film of chromium is plated over the nickel as a decorative and sealing coat.

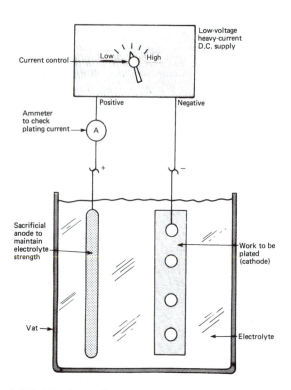

Figure 2.92 *Electroplating*

Electrolytic galvanising

This is the coating of low carbon steels with a layer of zinc. It is an electroplating process as described earlier. The metal deposited is zinc and the process is usually limited to flat and corrugated sheets. It is quicker and cheaper than hot-dip galvanising, but the coating is thinner.

Any corrosive attack on galvanised products eats away the zinc in preference to the iron. The zinc is said to be sacrificial. To prolong the life of galvanised sheeting it should be painted to protect the zinc itself from the atmosphere.

2.5.11 Surface coatings and treatments

A surface finish can be applied to many engineered products in order to improve their appearance, durability or corrosion resistance. We shall look at a variety of different finishes and how they are applied.

Grinding

A grinding wheel consists of abrasive particles bonded together. It does not 'rub' the metal away, it cuts the metal like any other cutting tool. Each abrasive particle is a cutting tooth. Imagine an abrasive wheel to be a milling cutter with thousands of teeth. Wheels are made in a variety of shapes and sizes. They are also available with a variety

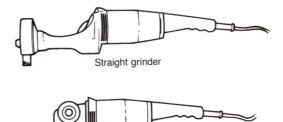
Straight grinder

Angle grinder

(a) Portable grinding machines

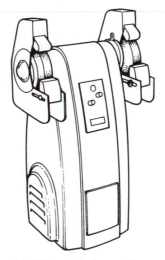

(b) Off-hand grinding machine

Figure 2.93 *Grinding machines*

of abrasive particle materials and a variety of bonds. It is essential to choose the correct wheel for any given job.

Figure 2.93(a) shows two types of electrically powered portable grinding machines used for dressing welds and for fettling castings. Care must be taken in its use and the operator should wear a suitable grade of protective goggles and a filter type respirator.

Figure 2.93(b) shows a double-ended, off-hand tool grinder. This is used for sharpening drills and lathe tools and other small tools and marking out instruments. It is essential to check that the guard is in place and that the visor and tool rest are correctly adjusted before commencing to use the machine. Grinding wheels can only be changed by a trained and certificated person.

Polishing

Polishing produces an even better finish than grinding but only removes the smallest amounts of metal. It only produces a smooth and shiny surface finish, the geometry of the surface is uncontrolled.

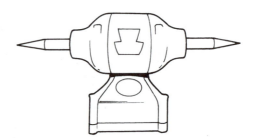

Figure 2.94 *A typical polishing lathe*

Polishing is used to produce decorative finishes, to improve fluid flow through the manifolds of racing engines, and to remove machining marks from surfaces that cannot be precision ground. This is done to reduce the risk of fatigue failure in highly stressed components.

Figure 2.94 shows a typical polishing lathe. It consists of an electric motor with a double-ended extended spindle. At each end of the spindle is a tapered thread onto which you screw the polishing mops. The mops may be made up from discs of leather (basils) or discs of cloth (calico mops). Polishing compound in the form of 'sticks' is pressed against the mops to impregnate them with the abrasive.

The components to be polished are held against the rapidly spinning mops by hand. As the mops are soft and flexible they will follow the contours of complex-shaped components. It is essential that dust extraction equipment is fitted to the polishing lathe and that the operator wears eye protection and a filter type dust mask. The process should only be carried out by a skilled polisher or under close supervision.

Electroplating
Electroplating has already been described when we discussed chemical treatments. Electroplating is the coating of metal components with another metal that is more decorative and/or corrosion resistant. Hot-dip galvanising coats low carbon steels with zinc without using an electroplating process.

Hot-dip galvanising
Hot-dip galvanising is the original process used for zinc coating buckets, animal feeding troughs and other farming accessories. It is also used for galvanised sheeting. The work to be coated is chemically cleaned, fluxed and dipped into the molten zinc. This forms a coating on the work. A small percentage of aluminium is added to the zinc to give the traditional bright finish. The molten zinc also seals any cut edges and joints in the work and renders them fluid tight. Metal components may also be coated with non-metallic surfaces.

Oxidising (oil blueing)
Steel components have a natural oxide film due to their reaction with atmospheric oxygen. This film can be thickened and enhanced by heating the steel component until it takes on a dark blue colour. Then

immediately dip the component into oil to seal the oxide film. This process does not work if there is any residual mill scale on the metal surfaces.

Chemical blacking

Alternatively, an even more corrosion-resistant oxide film can be applied to steel components by chemical blacking. The components are cleaned and degreased. They are then immersed in the oxidising chemical solution until the required film thickness has been achieved. Finally the treated components are rinsed, dewatered and oiled. Again, the process only works on bright surfaces.

Plastic coating

Plastic coatings can be both functional, corrosion resistant and decorative. The wide range of plastic materials available in a wide variety of colours and finishes provides a designer with the means of achieving one or more of the following:

- abrasion resistance
- cushioning effects with coatings up to 6 mm thick
- electrical and thermal insulation
- flexibility over a wide range of temperatures
- non-stick properties (Teflon PTFE coatings)
- permanent protection against weathering and atmospheric pollution, resulting in reduced maintenance costs
- resistance to corrosion by a wide range of chemicals
- the covering of welds and the sealing of porous castings.

To ensure success, the surfaces of the work to be plasticised must be physically and chemically clean and free from oils and greases. The surfaces to be plasticised must not have been plated, galvanised or oxide treated.

Fluidised bed dipping

Finely powdered plastic particles are suspended in a current of air in a fluidising bath as shown in Figure 2.95. The powder continually bubbles up and falls back and looks as though it is boiling. It offers no resistance to the work to be immersed in it. The work is preheated and immersed in the powder. A layer of powder melts onto the surface of the metal to form a homogeneous layer.

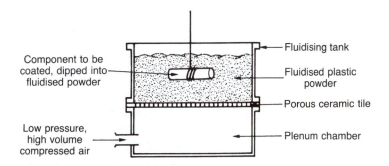

Figure 2.95 *Fluidised bed dipping*

Liquid plastisol dipping

This process is limited to PVC coating. A *plastisol* is a resin powder suspended in a plastisol and no dangerous solvents are present. The preheated work is suspended in the PVC plastisol until the required thickness of coating has adhered to the metal surface.

Painting

Painting is used to provide a decorative and corrosion-resistant coating for metal surfaces. It is the easiest and cheapest means of coating that can be applied with any degree of permanence. A paint consists of three components:

- *Pigment.* The finely powered pigment provides the paint with its opacity and colour.

- *Vehicle.* This is a film-forming liquid or binder in a volatile solvent. This binder is a natural or synthetic resinous material. When dry (set) it must be flexible, adhere strongly to the surface being painted, corrosion resistant and durable.

- *Solvent (thinner).* This controls the consistency of the paint and its application. It forms no part of the final paint film as it totally evaporates. As it evaporates it increases the concentration of catalyst in the 'vehicle' causing it to change chemically and set.

A paint system consists of the following components:

- *Primer.* This is designed to adhere strongly to the surface being painted and to provide a key for the subsequent coats. It also contains anti-corrosion compounds.

- *Putty or filler.* This is used mainly on castings to fill up and repair blemishes. It provides a smooth surface for subsequent paint coats.

- *Undercoat.* This builds up the thickness of the paint film. To produce a smooth finish, more than one undercoat is used with careful rubbing down between each coat. It also gives richness and opacity to the colour.

- *Top coat.* This coat is decorative and abrasion resistant. It also seals the paint film with a waterproof membrane. Modern top coats are usually based on acrylic resins or polyurethane rubbers.

There are four main groups of paint:

- *Group 1.* The vehicle is polymerised (see thermosetting plastics) into a solid film by reaction with atmospheric oxygen. Paints that set naturally in this way include traditional linseed oil based paints, oleo-resinous paints, and modern general purpose air drying paints based on modified alkyd resins.

- *Group 2.* This group of paints is based on amino-alkyd resins that do not set at room temperatures but they have to be *stoved* at 110–150°C. When set these paints are much tougher than group 1 air-drying paints.

- *Group 3.* These are the 'two-pack' paints. Polymerisation starts to occur as soon as the catalyst is mixed with the paint immediately before use. Modern 'one-pack' versions of these paints have the catalyst diluted with a volatile solvent as mentioned earlier. The solvent evaporates after the paint has been spread and, when the concentration of the catalyst reaches a critical level, polymerisation takes place and the paint sets.

- *Group 4.* These paints dry by evaporation of the solvent and no polymerisation occurs. An example is the cellulose paint used widely at one time for spray painting motor car body panels. Lacquers also belong to this group but differ from all other paints in that dyes are used as the colorant instead of pigments.

Paints may be applied by brushing, spraying or dipping. Whatever method is used, great care must be taken to ensure adequate ventilation. Not only can the solvents produce narcotic effects, but inhaling dried particles and liquid droplets of paint can cause serious respiratory diseases. The appropriate protective clothing, goggles and face masks must be used. Paints are also highly flammable and the local fire-prevention officer must be consulted over their storage and use. On no account can smoking be tolerated anywhere near the storage or use of paints.

Test your knowledge 2.67

State a suitable coating medium and describe its application process for each of the following engineered components:

(a) painting refrigerator body panels

(b) painting pressed steel angle brackets

(c) plastic cladding metal tubing for a bathroom towel rail.

Activity 2.16

A manufacturer of model boat kits has asked you to produce a single page instruction sheet on 'How to paint finish your kit'. Produce a word processed instruction sheet and illustrate it using appropriate sketches and drawings. Include a section headed 'Safety' and add the finished instruction sheet to your portfolio.

2.6 Health and safety

The Health and Safety at Work Act 1974 is based on principles that are fundamentally different from any previous health and safety legislation. The underlying reasoning behind the Act was the need to foster a much greater awareness of the problems which surround health and safety matters and, in particular, a much greater involvement of those who are, or who should be, concerned with improvements in occupational health and safety. Consequently, the Act seeks to promote greater personal involvement coupled with the emphasis on individual responsibility and accountability.

You need to be aware that the Health and Safety at Work Act applies to *people*, not to premises. The Act covers all employees in all employment situations. The precise nature of the work is not relevant, neither is its location. The Act also requires employers to take account of the fact that other persons, not just those that are directly employed, may be affected by work activities. The Act also places certain obligations on those who manufacture, design, import or supply articles or materials for use at work to ensure that these can be used in safety and do not constitute a risk to health.

> **Key point**
>
> The Health and Safety at Work etc., Act of 1974 makes both the employer and the employee (you) equally responsible for safety. Both can be prosecuted for violations of safety regulations and procedures.

> **Key point**
>
> It is your legal responsibility to take reasonable care for your own health and safety. The law expects you to act in a responsible manner so as not to endanger yourself or other workers or the general public.

> **Key point**
>
> It is an offence under the Health and Safety at Work act to misuse or interfere with equipment provided for your health and safety or the health and safety of others.

2.6.1 Causes of accidents

Human carelessness

Most accidents are caused by human carelessness. This can range from 'couldn't care less' and 'macho attitudes', to the deliberate disregard of safety regulations and codes of practice. Carelessness can also result from fatigue and ill-health and these, in turn, can result from a poor working environment.

Personal habits

Personal habits such as alcohol and drug abuse can render workers a hazard not only to themselves but also to other workers. Fatigue due to a second job (or 'moonlighting') can also be a considerable hazard, particularly when operating machines. Smoking in prohibited areas where flammable substances are used and stored can cause fatal accidents involving explosions and fire.

Supervision and training

Another cause of accidents is lack of training or poor quality training. Lack of supervision can also lead to accidents if it leads to safety procedures being disregarded.

Environment

Unguarded and badly maintained plant and equipment are obvious causes of injury. However, the most common causes of accidents are falls on slippery floors, poorly maintained stairways, scaffolding and obstructed passageways in overcrowded workplaces. Noise, bad lighting, and inadequate ventilation can lead to fatigue, ill-health and carelessness. Dirty surroundings and inadequate toilet and washing facilities can lead to a lowering of personal hygiene standards.

2.6.2 Accident prevention

Elimination of hazards

The workplace should be tidy with clearly defined passageways. It should be well lit and ventilated. It should have a well-maintained non-slip flooring. Noise should be kept down to acceptable levels. Hazardous processes should be replaced with less dangerous and more environmentally acceptable alternatives. For example, asbestos clutch and brake linings should be replaced with safer materials.

Guards

Rotating machinery, drive belts and rotating cutters must be securely fenced to prevent accidental contact. Some machines have interlocked guards. These are guards coupled to the machine drive in such a way that the machine cannot be operated when the guard is open for loading and unloading the work. All guards must be set, checked and maintained by qualified and certificated staff. They must not be

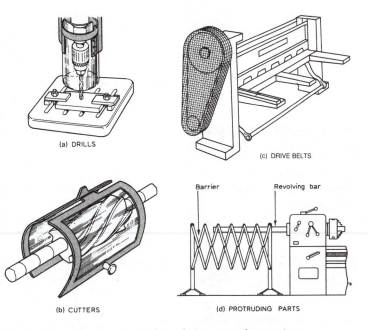

(a) DRILLS

(c) DRIVE BELTS

(b) CUTTERS

(d) PROTRUDING PARTS

Figure 2.96 *Some examples of the use of guards*

removed or tampered with by operators. Some examples of guards are shown in Figure 2.96.

Maintenance
Machines and equipment must be regularly serviced and maintained by trained fitters. This not only reduces the chance of a major breakdown leading to loss of production, it lessens the chance of a major accident caused by a plant failure. Equally important is attention to such details as regularly checking the stocking and location of First Aid cabinets and regularly checking both the condition and location of fire extinguishers. All these checks must be logged.

Personal protection
Suitable and unsuitable working clothing is shown in Figure 2.97.

Some processes and working conditions demand even greater protection, such as safety helmets, earmuffs, respirators and eye protection worn singly or in combination. Such protective clothing must be provided by the employer when a process demands its use. Employees must, by law, make use of such equipment. Some examples are shown in Figure 2.98.

Safety education
This is important in producing positive attitudes towards safe working practices and habits. Warning notices and instructional posters should be displayed in prominent positions and in as many ethnic languages as necessary. Information, education and training should be provided in all aspects of health and safety. For example process

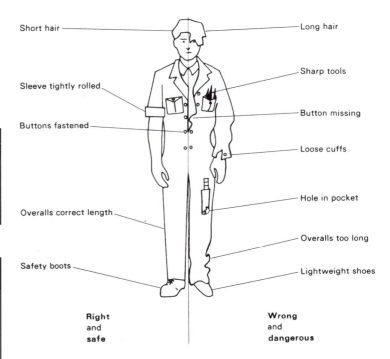

Figure 2.97 *Suitable and unsuitable clothing*

training, personal hygiene, first aid and fire procedures. Regular fire drills must be carried out to ensure that the premises can be evacuated quickly, safely and without panic.

Personal attitudes

It is important that everyone adopts a positive attitude towards safety. Not only your own safety but the safety of your workmates and the general public. 'Skylarking' and throwing things about in the workplace or on site cannot be allowed. Any distraction that causes lack of concentration can lead to serious and sometimes even fatal accidents.

Housekeeping

A sign of a good worker is a clean and tidy working area. Only the minimum of tools for the job should be laid out at any one time. These tools should be laid out in a tidy and logical manner so that they immediately fall to hand. Tools not immediately required should be cleaned and properly stored away. All hand tools should be regularly checked and kept in good condition. Spillages, either on the workbench or on the floor should be cleaned up immediately.

2.6.3 Electrical hazards

Electrical equipment is potentially dangerous. The hazards that are present in electrical and electronic equipment can include electric shock, fire and/or smoke fumes due to the overheating of cables and equipment. Explosions set off by sparks when using unsuitable equipment when flammable vapours and gasses are present.

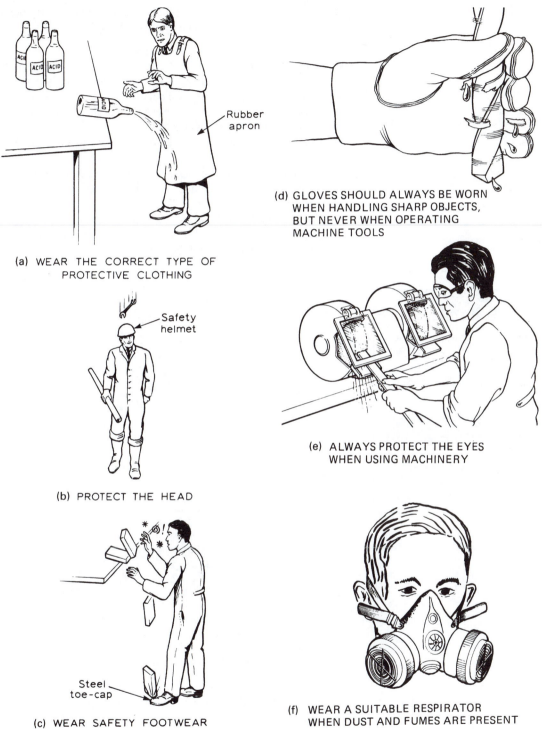

(a) WEAR THE CORRECT TYPE OF PROTECTIVE CLOTHING

Rubber apron

(b) PROTECT THE HEAD

Safety helmet

(c) WEAR SAFETY FOOTWEAR

Steel toe-cap

(d) GLOVES SHOULD ALWAYS BE WORN WHEN HANDLING SHARP OBJECTS, BUT NEVER WHEN OPERATING MACHINE TOOLS

(e) ALWAYS PROTECT THE EYES WHEN USING MACHINERY

(f) WEAR A SUITABLE RESPIRATOR WHEN DUST AND FUMES ARE PRESENT

Figure 2.98 *Safety equipment and clothing*

Before using any electrical equipment it is advisable to carry out a number of visual checks as shown in Figure 2.99(a):

- Check that the cable is not damaged or frayed.
- Check that both ends of the cable are secured in the cord grips of the plug or appliance and that none of the conductors is visible.
- Check that the plug is in good condition and not cracked.
- Check that the voltage and power rating of the equipment is suitable for the supply available.
- If low voltage equipment is being used check that a suitable transformer is available.
- Check that, whatever the voltage rating of the equipment, it is connected to the supply through a circuit breaker containing a residual current detector (RCD).
- Check that all metal clad electrical equipment has a properly connected earth lead and is fitted with a properly connected three-pin plug as shown in Figure 2.99(b).

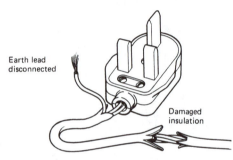

(a) EXAMINE PLUGS DAILY

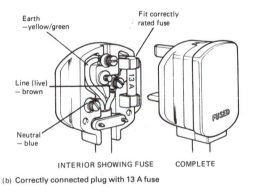

(b) Correctly connected plug with 13 A fuse

Figure 2.99 *Visual checks on electric plugs and cables*

Test your knowledge 2.70

State THREE causes of electrical accidents and suggest how they may be prevented.

Test your knowledge 2.71

List the checks you would make before using a portable mainsoperated electric power tool.

All exposed metalwork of electrically powered or operated equipment must be earthed to prevent electric shocks (see also battery charger project). Figure 2.100 shows the two ways in which a person may receive an electric shock. In Figure 2.100(a) the person is receiving a shock by holding both the live and neutral conductors so that the electric current can flow through his or her body. The neutral conductor is connected to earth, so it is equally possible to receive a shock via an earth path when holding a live conductor as shown in Figure 2.100(b).

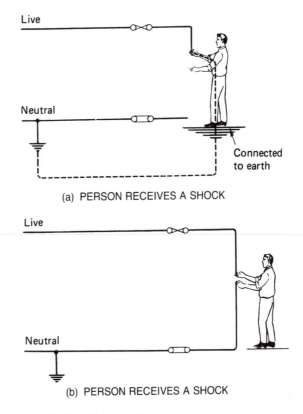

(a) PERSON RECEIVES A SHOCK

(b) PERSON RECEIVES A SHOCK

Figure 2.100 *Shock hazards*

It is unlikely you would be so foolish as to deliberately touch a live conductor, but you might come into contact with one accidentally.

For example, the portable electric drill shown in Figure 2.101(a) has a metal casing but no earth. The live conductor has fractured within the machine and is touching the metal casing. The operator cannot see this and would receive a serious or even fatal shock. The fault current would flow through the body of the user via the earth path to neutral.

Figure 2.101(b) shows the effect of the same fault in a properly earthed machine. The fault current would take the path of least resistance and would flow to earth via the earth wire. The operator would be unharmed. Electrical equipment must be regularly inspected, tested, repaired and maintained by qualified electricians. Note that 'double insulated' equipment does not need to be earthed.

It is important to remember that bodily contact with mains or high-voltage circuits can be lethal. The most severe path for electric current within the body (i.e. the one which is most likely to stop the heart) is that which exists from one hand to the other. The hand-to-foot path is also dangerous but somewhat less dangerous than the hand-to-hand path.

Voltages in many items of electronic equipment, including all items that derive their power from the a.c. mains supply, are at a level which can cause sufficient current flow in the body to disrupt normal operation of the heart. The severity of such a shock will depend upon

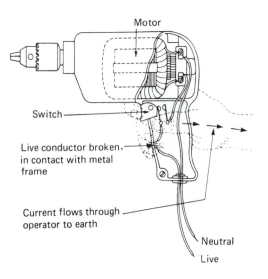

(a) ELECTRIC DRILL NOT EARTHED

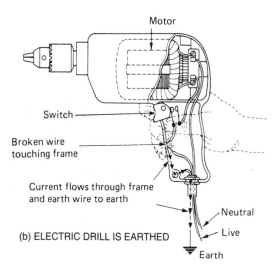

(b) ELECTRIC DRILL IS EARTHED

Figure 2.101 *Electric drill earthing to prevent shock hazard*

a number of factors including the magnitude of the current, whether it is alternating current (a.c.) or direct current (d.c.), and its precise path through the body. The magnitude of the current depends upon the voltage which is applied and the resistance of the body. The electrical energy developed in the body will depend upon the time for which the current flows; the duration of contact is also crucial in determining the eventual physiological effects of the shock. As a rough guide, and assuming that the voltage applied is from the 250 V a.c. 50 Hz mains supply, the effects shown in Table 2.14 are typical. It is important to realise that these figures are only quoted as a guide. In particular, you should note that there have been cases of lethal shocks resulting from contact with much lower voltages and at quite small values of current. Furthermore, an electric shock is often accompanied

Current	Physiological Effect
Less than 1 mA	Not usually noticeable
1–2 mA	A slight tingle may be felt
2–4 mA	Mild shock felt
4–10 mA	Serious shock (painful)
10–20 mA	Nerve paralysis may occur (unable to let go)
20–50 mA	Loss of consciousness (breathing may stop)
More than 50 mA	Heart failure

Table 2.14   *Effect of electric current on the human body*

by burns to the skin at the point of contact. Such burns may be extensive and deep even when there may be little visible external damage to the skin. Burns may be particularly severe when relatively high voltages and currents are encountered.

In general *any potential in excess of 50 V* should be considered dangerous. Lesser potential may, under unusual circumstances, also be dangerous. As such, it is wise to get into the habit of treating all electrical and electronic circuits with great care and avoid bodily contact at all times.

Residual current circuit breakers (RCCB), RCD, or earth leakage circuit breakers (ELCB) can provide a very worthwhile safety measure which can be instrumental in very significantly reducing the risk of electric shock. An RCCB senses the imbalance of current that occurs whenever current is returned to the a.c. mains supply via the earth rather than the neutral wire. This current may result from equipment failure (e.g. insulation breakdown in a mains transformer) or from bodily contact with the live (line) wire. Most RCCBs will trip at currents of about 30 mA within a time interval of 30 ms, sufficient to ensure that heart failure does not occur. RCCBs are not expensive and are available in various forms (ideally they should be wired into the mains circuit to your workshop so that all of the mains outlets are protected).

The following procedure should be adopted in the event of electric shock:

- Switch off the supply of current if this can be done quickly.
- If you cannot switch off the supply, do not touch the person's body with your bare hands (human flesh is a conductor and you would also receive a severe shock) but drag the affected person clear using insulating material such as dry clothing, a dry sack, or any plastic material that may be handy.
- If the affected person has stopped breathing commence artificial respiration immediately. Don't wait for help to come or go to seek for help.
- If the affected person's pulse has stopped, heart massage will also be required. If you are not confident to do this it is important to find another person to help who has had First Aid training.

Test your knowledge 2.72

State the typical operating current for an RCCB.

Test your knowledge 2.73

List the checks you would make before using a portable electric power tool.

Activity 2.17

Produce an A3 poster that can be displayed in your school or college workshop with information on how to deal with electric shock. Make sure that the poster is easy to read and understand. Add your finished poster to your portfolio.

2.6.4 Fire hazards

Fire is the rapid oxidation (burning) of flammable materials. For a fire to start, the following are required:

- a supply of flammable materials
- a supply of air (oxygen)
- a heat source.

Once the fire has started, the removal of one or more of the above will result in the fire going out.

Fire prevention

Fire prevention is largely 'good housekeeping'. The workplace should be kept clean and tidy. Rubbish should not be allowed to accumulate in passages and disused storerooms. Oily rags and waste materials should be put in metal bins fitted with airtight lids. Plant, machinery and heating equipment should be regularly inspected, as should fire alarm and smoke detector systems. You should know how to give the alarm.

Electrical installations, alterations and repairs must only be carried out by qualified electricians and must comply with the current IEE Regulations. Smoking must be banned wherever flammable substances are used or stored. The advice of the fire prevention officer of the local brigade should be sought before flammable substances, bottled gases, cylinders of compressed gases, solvents and other flammable substances are brought on site.

Fire procedures

In the event of you discovering a fire, you should:

- Raise the alarm and call the fire service.

- Evacuate the premises. Regular fire drills must be held. Personnel must be familiar with normal and alternative escape routes. There must be assembly points and a roll call of personnel. A designated person must be allocated to each department or floor to ensure that evacuation is complete. There must be a central reporting point.

- Keep fire doors closed to prevent the spread of smoke. Smoke is the biggest cause of panic and accidents, particularly on staircases. Emergency exits must be kept unlocked and free from obstruction whenever the premises are in use. Lifts must not be used in the event of fire.

- Only attempt to contain the fire until the professional brigade arrives if there is no danger to yourself or others. Always make sure you have an unrestricted means of escape. Saving lives is more important than saving property.

Fire extinguishers
Figure 2.102 shows a fire hose and a range of pressurised water extinguishers. These can be identified by their shape and colour which is RED. They are for use on burning solids such as wood, paper, cloth, etc. They are UNSAFE on electrical equipment at all voltages.

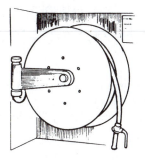

(a) HOSE REEL

(b) PRESSURISED WATER EXTINGUISHER

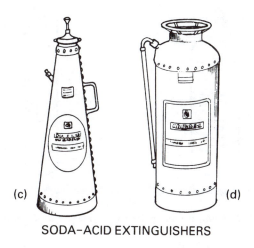

(c) (d)

SODA–ACID EXTINGUISHERS

Figure 2.102 *Various types of fire extinguisher*

Figure 2.103 shows two types of foam extinguisher. These can be identified by their shape and colour which is CREAM. They are for use on burning flammable liquids. They are UNSAFE on electrical equipment at all voltages.

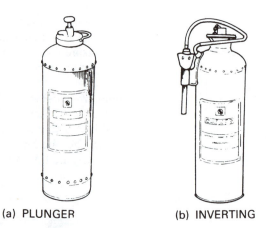

(a) PLUNGER   (b) INVERTING

Figure 2.103 *Foam extinguishers*

Figure 2.104 shows a variety of extinguishers that can be used on most fires and are safe for use on electrical equipment. Again, they can be identified by their shapes and colours:

- Dry powder extinguishers are coloured BLUE and are safe up to 1000 V.
- Carbon dioxide (CO_2) extinguishers are coloured BLACK and are safe at high voltages.
- Vaporising liquid extinguishers are coloured GREEN and are safe at high voltages.

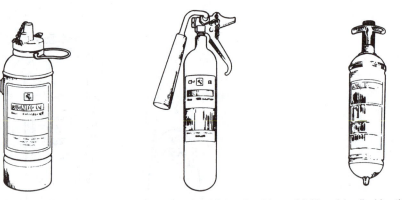

(a) Dry powder extinguisher   (b) Carbon dioxide (CO_2) extinguisher   (c) Vaporising liquid extinguisher

Figure 2.104 *Extinguishers for use on electrical equipment*

These latter two extinguishers act by replacing the air with an atmosphere free from oxygen. They are no good in draughts which would blow the vapour or gas away. Remember, if the fire cannot breathe neither can any living creature. Evacuate all living creatures before using one of these types of extinguisher. When using this type of extinguisher, keep backing away from the gas towards fresh air, otherwise it will put you out as well as the fire!

Figure 2.105 *A fire blanket*

Figure 2.105 shows a fire blanket. Fire blankets are woven from fire-resistant synthetic fibres and are used to smother fires. The old-fashioned blankets made from asbestos must NOT be used. The blanket is pulled from its container and spread over the fire to exclude the air necessary to keep the fire burning. They are suitable for use in kitchens, in workshops, and in laboratories. They are also used where a person's clothing is on fire, by rolling the person and the burning clothing up in the blanket to smother the fire. Do not cover the person's face.

Test your knowledge 2.74

State which type of fire extinguisher you would use in each of the following cases:

(a) Paper burning in an office waste bin.
(b) A pan of fat burning in the kitchen of the works canteen.
(c) A fire in a mains voltage electrical machine.

Test your knowledge 2.75

A fire breaks out near to a store for paints, paint thinners and bottled gases. What action should be taken and in what order?

Activity 2.18

Consult the Health and Safety at Work Act and answer the following questions:

(a) What is an improvement notice?
(b) What is a prohibition notice?
(c) Who issues such notices?
(d) Who can be prosecuted under the Act?

Present your findings in the form of a brief word processed fact sheet and add the report to your portfolio.

Activity 2.19

Investigate the construction of one of the lathes in your workshop and describe how the guard over the end train gears (change wheels) is interlocked so that the lathe cannot operate with the guard open. Illustrate your findings with a labelled sketch and add this to your portfolio.

Design an A3 poster that can be placed in your school or college workshop that will provide information on the types of fire extinguisher that can be safely used on different types of fire. Your poster should be designed so that it can be understood by people with only a very limited command of the English language.

Key point

Risk assessment takes into account the probability of an accident occurring as well as its likely severity.

2.6.5 Risk assessment

Risk assessment is a technique of evaluating not just the likelihood of an event occurring, but also what the outcome will be in terms of injury, loss, damage or harm. The level of risk is determined by the probability of an event as well as the likely severity of its outcome. The evaluation of the risk involved in a given process or activity is usually based on the following questions:

1. What is the hazard?
2. How likely is it to occur?
3. Under what circumstances might it occur?
4. What controls are in place to prevent it occurring?
5. What is the likely outcome if it does occur?

The process of identifying hazards can be a little daunting if you have never done anything like this before. To make things a little more manageable, you can divide the task into a number of smaller stages by answering questions like:

What	– processes are being used?
	– plant, materials and tools are being used?
	– is the effect on people/plant/other activities?
	– are the statutory requirements?
Who	– is doing the job? (and have they been trained?)
	– is exposed to risk?
	– might also be affected?
	– is supervising/monitoring the process?
Where	– is the process/activity performed?
	– is the process documented?
	– do waste materials go?
When	– is the process carried out?
	– were the safety procedures last reviewed/updated?
Why	– is the process being performed?
	– is the risk not being controlled?
	– is exposure not controlled at source?
How	– can the hazard be controlled?
	– can an accident occur?
	– may people be affected?

In order to answer all of these questions you need to observe the activity that is being performed as well as the circumstances in which it is being carried out. Further hazards may be identified by study of accident/incident reports, insurers' reports, inspection reports and other specialist survey reports.

Activity 2.21

Working in groups of three to four, use the following checklist to help you carry out a detailed and meaningful risk assessment of the electrical/electronic workshop in your school/college. You should copy the list and write brief notes against each question. Note that your risk assessment must be based on the *activities* that are carried out in the workshop so it is a good idea to start off by finding out *who* uses the workshop and *what* it is used for! Present your findings in the form of a brief presentation (no more than 10 min) using appropriate visual aids. The following questions should help you to get started:

1. Where is the electrical circuit breaker? (It should be in a prominent and immediately accessible position!)

2. Has a residual current circuit breaker (RCCB) been fitted and, if so, has the RCCB been tested lately?

3. Are all of the electrical appliances in a safe condition?

4. Does each item of test equipment have a mains lead that is in a safe condition?

5. Does each item of test equipment have a mains plug that is in a safe condition?

6. Is each item of test equipment correctly fused?

7. Have portable items of electrical equipment been tested in accordance with Portable Appliance Testing (PAT) regulations? If so, when was the last PAT test carried out?

8. Is the soldering equipment safe? Are soldering irons fitted with heatproof leads?

9. Is there any provision for solder fume extraction?

10. Are safety glasses available and what condition are they in?

11. Is the lighting adequate?

12. Are the safety exits properly marked and unobstructed?

13. Are appropriate fire extinguishers available and when were they last tested?

14. Is there appropriate safety information on display?

15. Is there a First Aid kit available?

2.6.6 First Aid

A knowledge of First Aid can and does save lives. If you arrive on the scene of an emergency where someone has received an injury or is unconscious (and even if you have never received any First Aid training) you should take the following initial action:

1. Remain calm. You can be most useful in a situation if you act in a calm, purposeful way. Be assertive and make others aware that you are in control of the situation.

2. Assess the situation quickly and calmly. Ascertain what has happened and try to identify the nature of the injury.

   Always be watchful for your own personal safety and any dangers that might still be present and might further threaten the casualty (e.g., gas, fire, etc.).

3. Make the area safe. Protect the casualty from further danger. At this stage do not try to do too much. Expert help will usually not be too far away. If there is more than one casualty, try to assess the priorities.

4. Ensure that appropriate help has been summoned (this will usually involve making a 999 call for an ambulance).

Key point

A knowledge of basic First Aid can save lives so, if you get the opportunity to receive First Aid training, do take it!

2.7 Revision problems

1. Suggest a suitable material for each of the following applications:

(a) The body of a screwdriver
(b) The blade of a screwdriver.

2. Explain the meaning of each of the following properties of a metal:

(a) Ductility
(b) Malleability
(c) Elasticity
(d) Compressibility.

3. A hard steel ball is pressed into a metal specimen using a standard force. What property of the material is being tested?

4. A resistor is marked with the following coloured bands; BROWN, RED, ORANGE, GOLD. What is the marked value and tolerance of the resistor?

5. State the type of machine suitable for producing:

(a) A cylindrical metal component
(b) A slotted metal plate
(c) A chisel blade.

6. List the sequence of operations required to make a large diameter hole in a metal plate.

7. During manufacture of a carbon steel component it is heated and quenched. Name and explain the purpose of this process.

8. List the sequence of operations required to make a printed circuit from a copper clad laminated plastic board.

9. List the sequence of operations required when painting a casting.

10. Explain the purpose of carrying out a risk assessment.

11. Describe the procedure for dealing with someone who is unconscious and who may have received an electric shock.

12. Explain the purpose of an RCCB or RCD device.

13. State the colours used for the following types of fire extinguisher:

(a) Carbon dioxide (CO_2)
(b) Dry powder
(c) Foam
(d) Vapourising liquid
(e) Water.

14. Name the machine shown in Figure 2.106 and state the process that it would be used for.

15. What essential safety precaution should be observed when using the machine shown in Figure 2.106?

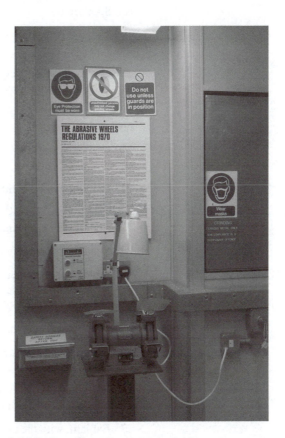

Figure 2.106 *See Questions 14 and 15*

Figure 2.107 *See Questions 17 and 18*

16. Explain, with the aid of sketches, the following engineering processes:

(a) Countersinking
(b) Counterboring
(c) Spot-facing.

17. Name the machine shown in Figure 2.107 and state the process that it is used for.

18. What essential safety precautions should be observed when using the machine shown in Figure 2.107?

19. It is necessary to toughen a carbon steel component. Explain how this is done.

20. Describe, with the aid of sketches, TWO methods of electronic circuit construction.

21. Explain, with the aid of a sketch, how a caliper gauge can be used to check the quality of a machined component.

22. State THREE advantages of adhesive bonding.

23. State the typical melting point range for:

(a) A typical soft solder
(b) A typical hard solder.

24. Describe, with the aid of sketches, TWO methods for joining two thin metal sheets together.

25. State TWO advantages of CNC machines over manually operated machines.

26. With the aid of a sketch, explain how a drift is used to remove a tapered shank drill from machine spindle.

27. Explain, with the aid of a sketch, how a crimped electrical connection is made.

Chapter 3 Application of technology

Summary

This chapter covers Unit 3 of the GCSE engineering curriculum. Technology has a major impact on the way that we design and manufacture an engineered product or deliver an engineering service and the use of technology – particularly new technology – is an exciting aspect of engineering. Technology provides us with new and more cost-effective solutions for problems that could not be easily solved using existing methods. In recent years the application of technology has revolutionised many traditional engineering processes, from cutting metal to manufacturing electronic circuit boards. At the same time, technology has given us the possibility of a whole range of new products from electric cars to consumer electronic equipment such as CD players and GPS receivers.

In this chapter you will look at the application of technology in engineering through a series of case studies. To help you understand real-life examples of the development and application of technology, each case study incorporates one or more activities for you to complete. The case studies also provide you with opportunities to investigate products and to develop skills in gathering and using information. In order to complete the case study activities you will need to make use of your school or college library as well as other information sources such as CD-ROMs and the World Wide Web.

Unlike the earlier chapters that were assessed through your portfolio, your achievement of the learning outcomes for this chapter will be assessed by means of a written examination.

3.1 Introduction

Imagine that you have the task of producing a printed circuit board that has to be drilled (using a small pillar drill) with 100, or so, 1 mm diameter holes through which component leads are to be inserted prior to soldering in place. In order that the component leads align with the copper pads on the track side of the circuit board, each of the holes has to be precisely positioned. Clearly, this task might take you some time but you would probably get there in the end!

Now imagine that you have to produce 1000 similar boards. Not only will this task take you a very long time but it would be highly likely that a significant number of boards would be rejected because the holes were not in the right place. What you need, of course, is to apply some technology to the solution and set up a machine to do the drilling for you!

At this point, it is worth stepping back a few years and considering how electronic circuits were manufactured 50 years ago and comparing this with the way they are manufactured today. Take a look at Figure 3.1 which shows the internal construction of a radio receiver designed and manufactured in the 1950s. Now look at Figure 3.2 which shows its modern equivalent. They do not seem to have a lot in common – even though both items of equipment essentially perform the same function. So, why the difference? The answer to this question

Figure 3.1 *Interior construction of a 1950s radio*

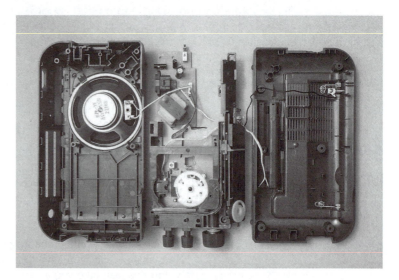

Figure 3.2 *Interior construction of a modern radio*

is simply that 'advances in technology (both that associated with the engineered product itself as well as the technology associated with its manufacture) have moved on!'

The important differences between the two radios shown in Figures 3.1 and 3.2 can be summarised in Table 3.1.

Feature	1950s radio	Modern radio
Enclosure	Wooden case	Moulded plastic case
Components	Six thermionic valves	One integrated circuit and diodes
Construction	Metal chassis	Two printed circuits
Wiring	Point-to-point soldered wires	
Power	30 W	Less than 3 W

Table 3.1 *Comparison of a 1950s radio with a current radio*

Key point

Continuous improvements in technology (both that which is directly associated with a product as well as that which is used in its manufacture) allow engineers to use increasingly cost-effective solutions.

Activity 3.1

Consider the way the telephone has changed in the last hundred years. Use your school or college library or other source, such as a CD-ROM or the Web, to obtain information on the development of the telephone from around 1900 to the present day, including mobile phones. Write a brief report and include sketches and illustrations showing how technology has changed not only the way the 'phone works but also how it is made. Do not forget to mention materials and processes in your report!

3.2 Manufacturing sectors

Before we move on, it is important to think about the various sectors that exist in engineering and manufacturing. In particular, you need to be able to relate an engineered product or engineering service (as well as the technology that it uses) to a particular sector, as follows:

Engineering
- engineering fabrication
- mechanical/automotive engineering
- electrical and electronic engineering, process control and telecommunications.

Manufacturing
- food and drink/biological and chemical
- printing and publishing/paper and board
- textiles and clothing.

Test your knowledge 3.1

Which TWO of the following engineered products belong to the mechanical engineering sector?

(a) Mobile phones
(b) Ready-to-cook meals
(c) CNC milling machines
(d) Portable CD players
(e) Lifting hoists
(f) Light aircraft.

Test your knowledge 3.2

Which TWO of the following engineered products belong to the electrical engineering sector?

(a) Metal cabinets
(b) Generators
(c) Gas welders
(d) Mobile phones
(e) Drinks cartons
(f) Transformers.

Key point

Engineered products and services are often attributed to the sector from which they originate.

Key point

Engineering sectors include aerospace, electrical and electronic, mechanical, telecommunications, automotive and manufacturing.

The engineering sectors, products and leading companies that you need to be particularly familiar with are as follows:

Aerospace
New passenger and military aircraft, satellites, space vehicles, missiles, etc. from companies such as British Aerospace, Westland and Rolls-Royce.

Electrical and electronic
Electric generators and motors, consumer electronic equipment (radio, TV, audio and video) power cables, computers, etc. produced by companies such as GEC, BICC and ICL.

Mechanical
Bearings, agricultural machinery, gas turbines, machine tools and the like from companies such as RHP, GKN and Rolls-Royce.

Telecommunications
Telephone, radio and data communications equipment, etc. from companies like Nokia, GEC, Plessey and British Telecom.

Within the *manufacturing sector* you need to be able to identify and investigate a variety of products as well as the engineering technology used in their manufacture. An example of this is the use of engineering in the printing industry.

Activity 3.2

Investigate the engineering industry in your area. Name at least three engineering companies and identify the engineering sectors in which they are active. Give ONE example of how EACH of these companies is making use of technology. Present your findings in the form of a brief class presentation using appropriate visual aids.

Activity 3.3

In recent years, the application of technology in automotive engineering has led to a number of significant improvements in motor vehicle design. Obtain literature from a local car dealer and identify THREE improvements that are incorporated in modern production vehicles that were unavailable in vehicles produced 20 years ago. At least ONE of these improvements should contribute to the safety of the driver and passengers. Present your findings in the form of a brief class presentation using appropriate visual aids.

Figure 3.3 shows the exploded view of an electric saw.

(a) To which engineering sector does this product belong?

(b) Give the part numbers of THREE components that are electrical and have been developed from electrical technology.

(c) Give the part numbers of THREE components that are mechanical and have been developed from mechanical technology.

(d) Give the part numbers of THREE components that are polymers (plastic) and have been developed from materials technology.

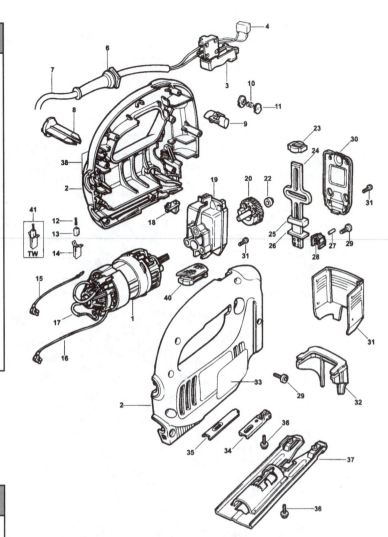

Figure 3.3 *Exploded view of an electric saw*

Figure 3.4 shows the exploded view of a VHF transceiver for use in a vehicle.

(a) To which engineering sector does this product belong?

(b) Give the part numbers of THREE components that are electrical and have been developed from electrical technology.

(c) Give the part numbers of THREE components that are mechanical and have been developed from mechanical technology.

(d) Give the part numbers of THREE components that are polymers (plastic) and have been developed from materials technology.

Activity 3.4

Several different types of computer printer (including ink jet and laser types) are currently available to home, small business and education users. Investigate one ink jet and one laser printer. Obtain a full specification and a user manual for each type (your tutor may be able to help you with this) and use this, together with other sources of information, to investigate the technology used for printing. Draw up a comparison chart which lists features such as print technology, media requirements, cost (and cost of consumables), operating lifetime, colour print capability, reliability, etc. Also investigate the impact that the print technology has on the design of the printer. Present your result in the form of a single A4 page 'fact sheet' entitled 'Print Technology for Home and Small Business Users'.

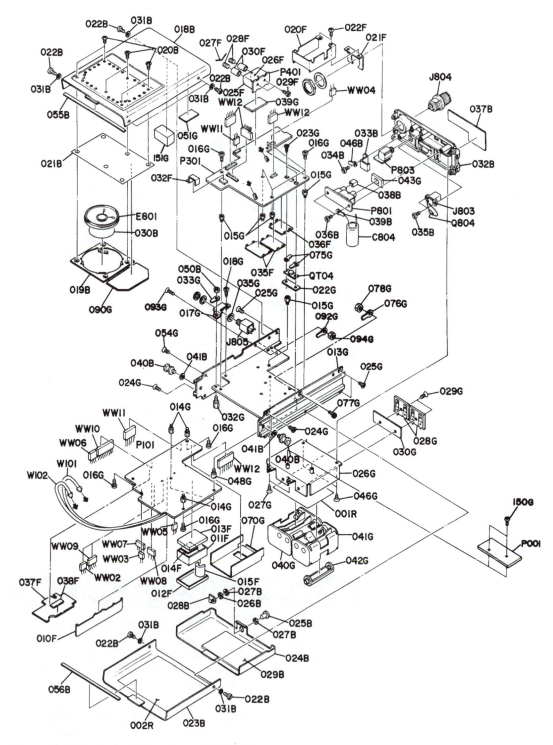

Figure 3.4 *Exploded view of a mobile VHF transceiver*

3.3 New technology

As we saw earlier when we looked at a simple domestic radio receiver, 'new technology' is replacing 'old technology' on an ongoing basis. Here are a few more examples of where developments in component, materials and manufacturing technology have resulted in major benefits in terms of a combination of improved cost effectiveness, better aesthetic features, enhanced functionality and increased reliability:

- transparent acrylic materials (offering both strength and lightness) replacing conventional glass
- light emitting diodes (LEDs) replacing conventional filament lamp indicators
- liquid crystal display screens replacing cathode ray tubes in computer displays
- composite materials, such as glass-reinforced plastic (GRP), replacing metal components in the bodies of high-performance vehicles.

As you learn about the use of new technology, you need to be able to relate it to a range of engineered products and services from across a number of sectors, just as you did before. Some particularly good examples of the use of new technology can be drawn from:

- the use of information and communications technology (ICT), including computer aided design (CAD), computer aided manufacture (CAM), communications technology and control technology

- the use of modern and smart materials and components, including polymers (plastics, adhesives and coating), metals and composites, microelectronic components (including microprocessors and memory devices used in computer and control technology)

- the use of systems and control technology to organise, monitor and control production, including process controllers, automation systems and programmable logic controllers (PLCs).

You also need to understand the impact of these modern technologies on:

- the range, types and availability of products
- the design and development of products
- the materials and components used.

Test your knowledge 3.5

Explain the meaning of each of the following abbreviations:

(a) CAD
(b) CAM
(c) GRP
(d) ICT
(e) PLC.

3.3.1 Information and communications technology

Information and communications technology (ICT) is not just about computers but about all aspects of accessing, processing and disseminating information. This revolution has come about as a result of the convergence of several different technologies, notably:

- computing (microprocessors, memories, magnetic and optical data storage devices, etc.)
- telecommunications (data communications, networking, optical fibres, etc.)

• software (optical character recognition, data transfer protocol, visually orientated programming languages, etc.).

The first case study in this section deals with databases whilst the second is devoted to the Internet and the World Wide Web. The final case study is about computer aided engineering (CAE).

Case study

Databases

A database is simply an organised collection of data. This data is usually organised into a number of records each of which contains a number of fields. Due to their size and complexity and the need to be able to quickly and easily search for information, a database is usually stored within a computer and a special program – a *database manager* or *database management system* (DBMS) – provides an interface between users and the data itself. The DBMS keeps track of where the information is stored and provides an index so that users can quickly and easily locate the information they require (see Figure 3.5).

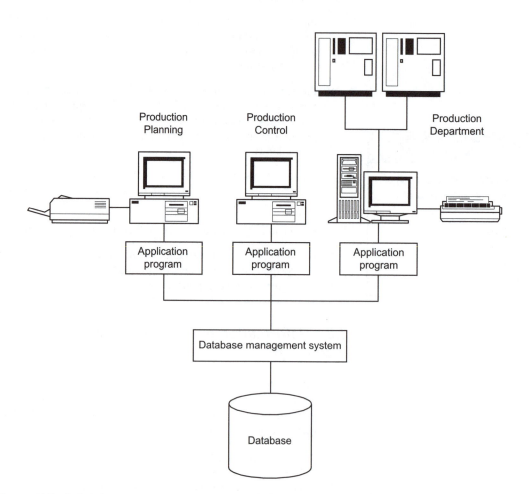

Figure 3.5 *A database management system (DBMS)*

The database manager will also allow users to search for related items. For example, a particular component may be used in a number of different products. The database will allow you to quickly identify each product that uses the component as well as the materials and processes that are used to produce it.

The structure of a simple database is shown in Figure 3.6. The database consists of a number of *records* arranged in the form of a table. Each record is divided into a number of *fields*. The fields contain different information but they all relate to a particular component. The fields are organised as follows:

Field 1 Key (or index number)
Field 2 Part number
Field 3 Type of part
Field 4 Description or finish of the part.

Take a look at Record 100 in the database. This shows the data held for a steel washer with a part number W7392. Record 101, on the other hand, is for a black 2 mm connector with a part number C1020.

In most engineering companies several databases are used including a product database, customer database and a spare parts database. Recently, however, there has been a trend towards integrating many of the databases within an engineering company into one large database. This database becomes central to all of the functions within the company. In effect, it becomes the 'glue' that holds all of the departments together.

The concept of a centralised manufacturing database is a very sound one because it ensures that every function within the company has access to the same data. By using a single database, all departments become aware of changes and modifications at the same time and there is less danger of data becoming out of date.

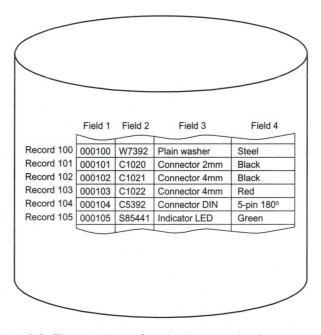

Test your knowledge 3.6

A database of parts is used in the manufacture of a range of motorcycles. Explain how this database might be organised and illustrate your answer by giving an example of a record for a front fork component.

	Field 1	Field 2	Field 3	Field 4
Record 100	000100	W7392	Plain washer	Steel
Record 101	000101	C1020	Connector 2mm	Black
Record 102	000102	C1021	Connector 4mm	Black
Record 103	000103	C1022	Connector 4mm	Red
Record 104	000104	C5392	Connector DIN	5-pin 180°
Record 105	000105	S85441	Indicator LED	Green

Figure 3.6 *The structure of a simple parts database*

Activity 3.5

Investigate the use of databases in your school or college. Working as part of a group, begin by finding out the name of the person who has overall responsibility for collecting and processing student data. Interview him or her and find out what information is held in the database and how it is organised. Also find out about the reports that are generated by the database and who has access to the information. Prepare a brief presentation to the rest of your class using appropriate handouts and visual aids.

Case study

The Internet and the World Wide Web

Although the terms *Web* and *Internet* are often used synonymously, they are actually two different things. The *Internet* is the global association of computers that carries data and makes the exchange of information possible. The *World Wide Web* is a subset of the Internet – a collection of inter-linked documents that work together using a specific Internet protocol called *hypertext transfer protocol* (HTTP). In other words, the Internet exists independently of the World Wide Web, but the World Wide Web cannot exist without the Net.

Key point

Every Web page on the Internet has its own unique address called a URL.

The World Wide Web began in March 1989, when Tim Berners-Lee of the European Particle Physics Laboratory at CERN (the European centre for nuclear research) proposed the project as a means to better communicate research ideas among members of a widespread organisation.

Web sites are made up of collections of Web pages. Web pages are written in *hypertext markup language* (HTML), which tells a *Web browser* (such as Netscape's Communicator or Microsoft's Internet Explorer) how to display the various elements of a Web page. Just by clicking on a *hyperlink*, you can be transported to a site on the other side of the world.

A set of unique addresses is used to distinguish the individual sites on the World Wide Web. An Internet Protocol (IP) address is a 4- to 12-digit number that identifies a specific computer connected to the Internet. The digits are organised in four groups of numbers (which can range from 0 to 255) separated by full stops. Depending on how an Internet Service Provider (ISP) assigns IP addresses, you may have one address all the time or a different address each time you connect.

Test your knowledge 3.7

Take a look at the engineering company's Web page shown in Figures 3.7 and 3.8.

(a) What is the name of the company?
(b) What is the specialism of the company?
(c) In what engineering sector does the company operate?
(d) What is the full URL for the company's Web page?
(e) How many hyperlinks are available to other pages from the company's Web page?

Every Web page on the Internet, and even each object that you see displayed on a Web page, has its own unique address, known as a *uniform resource locator* (URL). The URL tells a browser exactly where to go to find the page or object that it has to display. It is important to realise that, when a Web page is sent over the Internet, what is actually sent is not a complete page (as it might appear in a word processor) but a set of instructions that allow the page to be reconstructed by a browser.

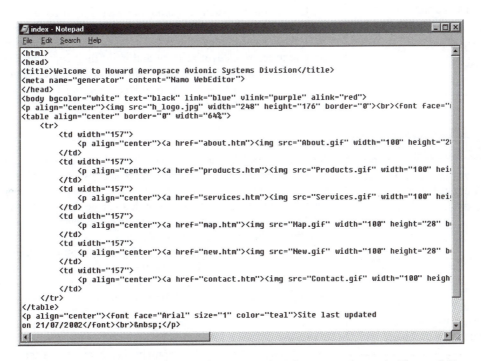

Figure 3.7 *A typical engineering company's Web page displayed in a Web browser*

```
<html>
<head>
<title>Welcome to Howard Aeropsace Avionic Systems Division</title>
<meta name="generator" content="Namo WebEditor">
</head>
<body bgcolor="white" text="black" link="blue" vlink="purple" alink="red">
<p align="center"><img src="h_logo.jpg" width="248" height="176" border="0"><br><font face="
<table align="center" border="0" width="64%">
    <tr>
        <td width="157">
            <p align="center"><a href="about.htm"><img src="About.gif" width="100" height="2
        </td>
        <td width="157">
            <p align="center"><a href="products.htm"><img src="Products.gif" width="100" hei
        </td>
        <td width="157">
            <p align="center"><a href="services.htm"><img src="Services.gif" width="100" hei
        </td>
        <td width="157">
            <p align="center"><a href="map.htm"><img src="Map.gif" width="100" height="28" b
        </td>
        <td width="157">
            <p align="center"><a href="new.htm"><img src="New.gif" width="100" height="28" b
        </td>
        <td width="157">
            <p align="center"><a href="contact.htm"><img src="Contact.gif" width="100" heigh
        </td>
    </tr>
</table>
<p align="center"><font face="Arial" size="1" color="teal">Site last updated
on 21/07/2002</font><br> </p>
```

Figure 3.8 *The HTML code responsible for generating the page shown in Figure 3.7*

E-mail

Many people now use the Internet for communicating with other people by e-mail. Whilst this provides a simple means of sending and receiving simple text messages, it is also possible to transmit one or more files along with an e-mail message. This facility makes it possible to transfer images, word-processed documents, spreadsheets and database files by using e-mail. In order to use e-mail you need to open an account with an e-mail service provider. Such an account can be Web based (in which case you can send and receive e-mail from anywhere) or may be specific to a particular school, college or company (in which case the service is only available from within the establishment).

Search sites and search engines

Being able to locate the information that you need from a vast number of sites scattered across the globe can be a daunting prospect. However, since this is a fairly common requirement, a special type of site, known as a *search site*, is available to help you with this task. There are three different types of search site on the Web: *search engines*, *Web directories* and *metasearch sites*.

Search engines such as Excite, HotBot and Google use automated software called *Web crawlers* or *spiders*. These programs move from Web site to Web site, logging each site title, URL, and at least some of its text content. The object is to hit millions of Web sites and to stay as current with them as possible. The result is a long list of Web sites placed in a database that users search by typing in a keyword or phrase.

Web directories such as Yahoo and Magellan offer an editorially selected, topically organised list of Web sites. To accomplish that goal, these sites employ editors to find new Web sites and work with programmers to categorise them and build their links into the site's index. To make things even easier, all the major search engine sites now have built-in topical search indexes, and most Web directories have added a keyword search.

Intranets

Intranets work like the Web (with browsers, Web servers and Web sites) but companies and other organisations use them internally. Companies use them because they let employees share corporate data, but they are cheaper and easier to manage than most private networks – no one needs any software more complicated or more expensive than a Web browser, for instance.

They also have the added benefit of giving employees access to the Web. Intranets are closed off from the rest of the Net by firewall software, which lets employees surf the Web but keeps all the data on internal Web servers hidden from those outside the company.

Extranets

One of the most recent developments has been that of the *Extranet*. Extranets are several Intranets linked together so that businesses can share information with their customers and suppliers. Consider, for example, the production of a European aircraft by four major

Key point

Newsgroups and bulletin boards allow engineers to keep in touch with one another and with the latest developments.

aerospace companies located in different European countries. They might connect their individual company Intranets (or parts of their Intranets) to each other, using private leased lines or even using the public Internet. The companies may also decide to set up a set of private *newsgroups* and *bulletin boards* with *discussion threads* so that employees from different companies can exchange ideas and share information. The ability to research, convey and share information helps to make the Internet indispensable to any engineering company!

Activity 3.6

Visit the Web site of Lansing Linde, UK, a manufacturer of forklift trucks. Investigate the company and view some of the gallery of photographs that show the development of the company's forklift trucks. Search the site for information on its latest range of electric forklift trucks capable of handling loads from 1000 to 8000 kg. Write a brief report describing these trucks and make specific reference to features that:

- save energy
- ensure smoother driving
- protect the driver
- improve visibility
- facilitate turning and manoeuvring
- ensure stability.

The URL of the Lansing Linde Web site is:

http://www.lll.co.uk

Activity 3.7

Use a search engine (such as Lycos or Google) to locate information about motorcycle manufacturers. Visit the first four sites displayed as a result of your search and note down the URL of each of these sites. Summarise the contents of each of the sites by writing a paragraph describing each site. Then rate each site on a scale of 1–10 on the basis of content, presentation and ease of use. Summarise your results in a table. Repeat the activity using a Web directory (such as Yahoo or Excite). Use the directory to navigate to four different sites giving details of motorcycle manufacturers. Once again, note down the URL of each site, summarise its contents and rate it on a scale of 1–10 (again presenting your results in the form of a table). Compare these two search methods.

Activity 3.8

If you have not already done so, set up a Web-based e-mail account in your own name. Note down all of the steps that you took to open the account including details of any electronic forms that you had to complete. Present your findings in the form of a brief article for your local paper on how to open and use an e-mail account. You should assume that the reader is non-technical.

Case study

Computer aided engineering

From your study of Chapter 1, you will already know a little about CAD however this is just one aspect of a bigger subject, computer aided engineering (CAE). CAE is about automating *all* of the stages that go into providing an engineered product or service.

When applied effectively, CAE ties all of the functions within an engineering company together. Within a true CAE environment, information (i.e. data) is passed from one computer aided process to another. This often involves computer simulation, computer aided drawing (drafting) and CAM.

The single term, CADCAM, was once used to describe the integration of CAD, drafting and manufacture. Another term, CIM (computer integrated manufacturing), is nowadays used to describe an environment in which computers are used as a common link that binds together the various different stages of manufacturing a product, from initial design and drawing to final product testing. Whilst all of these abbreviations can be confusing (particularly as some of them are used interchangeably) it is worth remembering that 'computer' appears in all of them. What we are really talking about is the application of computers within engineering. Nowadays, the boundaries between the strict disciplines of CAD and CAM are becoming increasingly blurred.

We have already said that CAD is often used to produce engineering drawings. Several different types of CAD are used in engineering and we have shown some examples in Figures 3.9–3.12.

CAM covers a number of more specialised applications of computers in engineering including CIM, manufacturing system modelling and simulation, systems integration, artificial intelligence applications in manufacturing control, CAD/CAM, robotics and metrology. CAE analysis can be conducted to investigate and predict mechanical, thermal and fatigue stress, fluid flow and heat transfer, and vibration/noise characteristics of design concepts to optimise final product performance. In addition, metal and plastic flow, solid modelling and variation simulation analysis are performed to examine the feasibility of manufacturing a particular part.

In addition, all of the machine tools within a particular manufacturing company may be directly linked to the CAE network. Indeed, most modern manufacturing plants rely heavily on CIM systems.

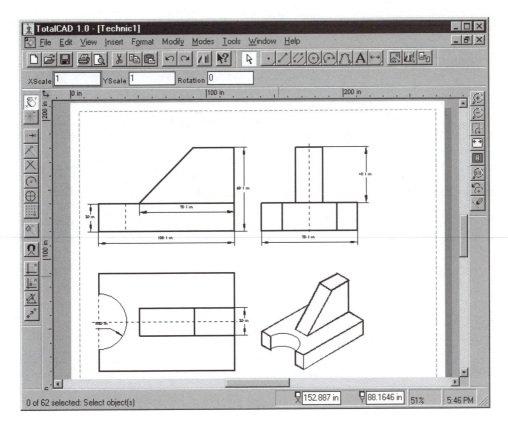

Figure 3.9 *A conventional engineering drawing produced by a CAD package*

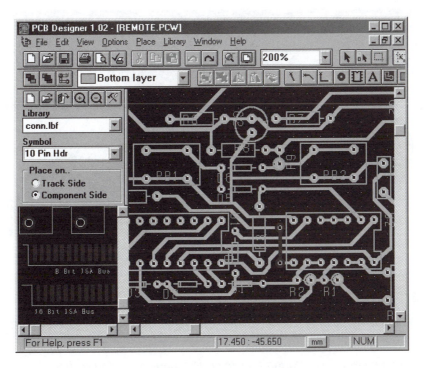

Figure 3.10 *Printed circuit board design is another excellent application for CAD*

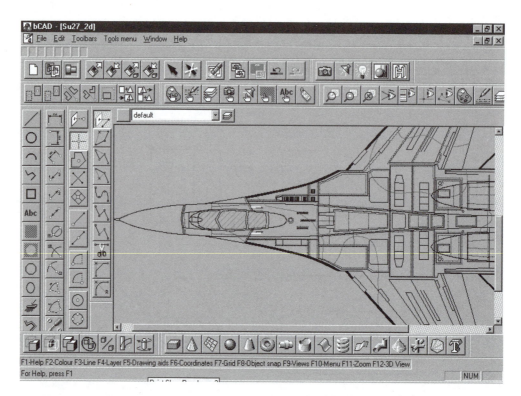

Figure 3.11 *Complex drawings can be produced prior to generating solid 3D views*

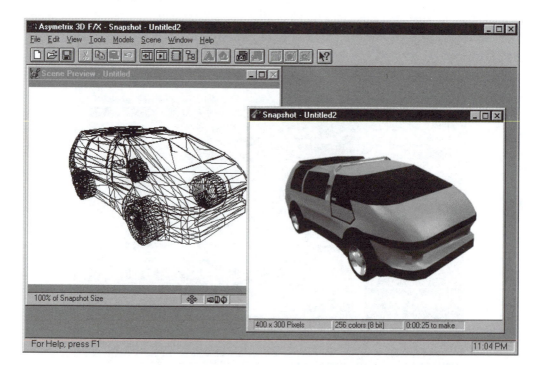

Figure 3.12 *A wire frame 3D drawing and its corresponding rendered view*

Some of the most advanced automated systems are employed by those industries that process petrol, gas, iron and steel. On a somewhat smaller scale, the manufacture of cars and trucks frequently involves computer-controlled robot devices. Similar industrial robots are used in a huge range of applications that involve assembly or manipulation of components.

Another development that has greatly affected manufacturing industries is the integration of engineering design and manufacturing into one continuous automated activity through the use of computers. The introduction of CIM has significantly increased productivity and reduced the time required to develop new products. When using a CIM system, an engineer develops the design of a component directly on the display screen of a computer. Information about the component and how it is to be manufactured is then passed from computer to computer within the integrated CAD/CAM system.

After the design has been tested and approved, the CIM system prepares instructions for computer numerically controlled (CNC) machine tools and places orders for the required materials and additional parts (such as nuts, bolts or adhesives). The CAD/CAM system allows an engineer (or, more likely, a team of engineers) to perform all the activities of engineering design by interacting with a computer system (invariably networked) before actually manufacturing the component in question using one or more CNC machines linked to the CAE system.

Test your knowledge 3.11

Explain the advantages of a CIM system.

Key point

The introduction of CIM has significantly increased productivity and reduced the time required to develop new products. When using a CIM system, an engineer develops the design of a component directly on the display screen of a computer.

Information about the component and how it is to be manufactured is then passed from computer to computer within the integrated CAD/CAM system.

Activity 3.9

Investigate a CNC machine in your school or college (if your school does not have such a machine they may be able to arrange for you to visit a local engineering manufacturer or a nearby Further Education college that has this equipment). Find out what the CNC machine is used for and how it is programmed. Write a brief report describing what you have seen and explain, in simple terms, how a CNC machine works. Illustrate your report with relevant sketches, diagrams and photographs.

Activity 3.10

Use the Internet to search for information about CNC milling machines. Obtain the specifications for THREE different CNC milling machines and present these in the form of a brief word-processed fact sheet.

Case study

GPS receivers

The Global Positioning System (or GPS) is a collection of satellites owned by the US Government that provides highly accurate, world-wide positioning and navigation information, 24 h a day. It is made up of 24 NAVSTAR GPS satellites that orbit 12,000 miles above the earth, constantly transmitting the precise time and their position in space. GPS receivers on (or near) the surface of the earth, listen in on the information received from 3 to 12 satellites and, from that, determine the precise location of the receiver, as well as how fast and in what direction it is moving.

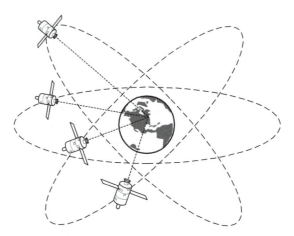

Figure 3.13 *GPS system using triangulation – a minimum of three satellites is required to determine position*

GPS uses the triangulation of signals from the satellites to determine locations on earth (see Figure 3.13). GPS satellites know their location in space and receivers can determine their distance from a satellite by using the travel time of a radio message from the satellite to the receiver. After calculating its relative position to at least three or four satellites, a GPS receiver can calculate its position using triangulation. GPS satellites have four highly accurate atomic clocks on board. They also have a database (sometimes referred to as an 'almanac') of the current and expected positions for all of the satellites that is frequently updated from earth. That way when a GPS receiver locates one satellite, it can download all satellite location information, and find the remaining needed satellites much more quickly.

Over the last several years, an increasing variety of affordable GPS receivers have been released for the average consumer. As the technology has improved, many additional features are added to these units, while the price and size continue to decrease.

Some of the more specialised GPS receivers currently available include:

- hand-held GPS receivers that have background maps
- GPS receivers fitted to cars and lorries (with integral databases of maps and road information)

- GPS receivers for large and small boats (including those integrated with other navigational equipment)
- aircraft GPS receivers with built-in airport information
- GPS receivers that combine with Internet access and/or e-mail into one unit.

GPS products have been developed for use for many commercial applications including surveying, map making, tracking systems, navigation, construction and mapping natural resources. The recreational use of GPS includes sea fishing, hiking, skiing and mountain walking.

Activity 3.11

Obtain the data sheets and/or technical specification for at least two different types of low-cost GPS receiver. Identify the market for these receivers and write a brief article for your local newspaper explaining what GPS can do for ordinary readers.

Test your knowledge 3.12

What is the minimum number of satellites within sight of a GPS receiver needed to fix a position? Explain your answer.

Activity 3.12

Visit a local car showroom and investigate the GPS systems that may be supplied with the latest models. Write a brief article for a car enthusiasts' magazine explaining, in simple terms, how GPS works and how it can benefit the car driver.

3.3.2 Systems and control technology

Control systems are used in many engineered products, including aircraft, cars, and consumer electronic equipment such as videos, CD and DVD players. Control systems invariably comprise a number of elements, components or sub-systems that are connected together in a particular way. The individual elements of the control system work together in order to make sure that the desired output is maintained. All control systems have at least one input and output.

A simple control system is shown in Figure 3.14. This system has a single input, the desired value (or *set point*) and a single output (the *controlled variable*). Inside the system there is a *controller* (often this is an electronic circuit which may be either digital or analogue – see later), and a *final control element* (often this is a motor or an actuator that causes something to move but it could also be a heater or some other form of *transducer* – see later). The *controlled process* is simply the name given to the overall process that produces the controlled variable.

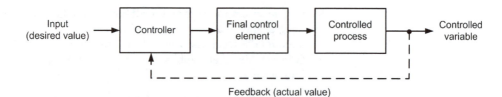

Figure 3.14 *A simple control system*

In most practical control systems *feedback* is included in order to sample the actual value and provide automatic correction. The feedback signal is applied to the controller which then compares it with the desired value and then makes adjustments accordingly. In case this is beginning to sound rather complicated it is worth taking a look at a simple example.

The angle of pitch attitude of an aircraft (basically this is about whether the nose of an aircraft is pointing upwards or downwards) is controlled by moving control surfaces called *elevators*. The pilot of an aircraft (whether it is a human pilot or an *autopilot*) needs to be able to control the aircraft pitch and a control system is used to do this.

In this case, a movement of the aircraft's joystick or control column operates an electric motor that drives a shaft and gearbox that in turn operates the elevator control surface. At the same time, the position of the elevator is detected and fed back to the pitch attitude controller, so that small adjustments can continually be made to maintain the desired attitude.

The control system uses three basic components:

- the pitch controller (the controller)
- an elevator actuator (the final control element)
- the controlled process (the adjustment of elevator angle and aircraft pitch).

Figure 3.15 shows the aircraft pitch attitude control system represented in both diagrammatic and block schematic forms.

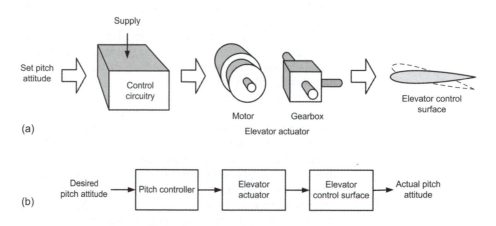

Figure 3.15 *The pitch control system shown in both diagrammatic and block schematic forms*

An important feature of many control systems is that operation is *automatic* so that, once set, the system is usually capable of operating with minimal human intervention. In the earlier example, when the autopilot is selected, adjustment of the elevator control surface and aircraft pitch becomes entirely automatic and the aircraft maintains the required attitude using feedback to correct for any disturbances or variations.

Analogue and digital systems

Control systems can be based on analogue or digital methods or they can use a mixture of the two. Analogue control involves the use of signals and quantities that are continuously variable. Within analogue control systems, signals are represented by voltages and currents that can take any value between two set limits. Figure 3.16 shows how a typical analogue signal varies with time.

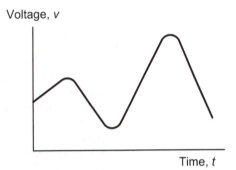

Figure 3.16 *A typical analogue signal*

Digital control, on the other hand, involves the use of signals and quantities that vary in discrete steps. Any values that fall between two adjacent steps must take one or other value as intermediate values are not allowed in a digital system. Figure 3.17 shows how a typical digital signal varies with time.

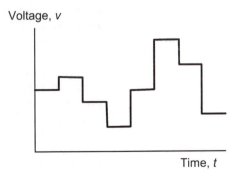

Figure 3.17 *A typical digital signal*

Sensors and transducers

Transducers are devices that convert energy in the form of sound, light, heat, etc. into an equivalent electrical signal, or vice versa. For example, a *loudspeaker* is a device that converts low-frequency electric current into sound. A *thermocouple*, on the other hand, is a device that converts temperature into voltage.

A *sensor* is simply a transducer that is used to generate an input signal to a control or measurement system. The signal produced by a sensor is an *electrical analogy* of a physical quantity, such as angular position, distance, velocity, acceleration, temperature, pressure, light level, etc. In a complex control system, the signals returned from several sensors, together with control inputs from the operator (where appropriate) will subsequently be used to determine the output from the system.

3.3.3 Materials and components

Ongoing developments in materials and components have made possible a number of exciting new products. There are three case studies in this section. The first deals with optical fibres that are increasingly being used to replace copper cables in telecommunications and computer networks. The second case study introduces a group of materials commonly referred to as plastics but more correctly called polymers. Due to their unique properties, these materials are now widely used in engineering. The third case study is devoted to a class of materials that has revolutionised the manufacture of boats, cars and aircraft. The composite materials are now widely used and are increasingly replacing fabricated metal parts. Integrated circuits (and microprocessors in particular) are introduced in the final case study in this section. Microprocessors form the basis of the microcontrollers and PLCs that we shall meet later in this chapter.

Case study   Optical fibres

An optical fibre is a long thin strand of very pure glass enclosed in an outer protective jacket. As the refractive index of the inner layer is larger than that of the outer layer, light travels along the fibre by means of total internal reflection at a speed of around 200 million m/s (approximately 2/3 of the speed of light).

In order to set up an optical data link, the optical fibre must be terminated at each end by means of a transmitter/receiver unit. A simple one-way fibre optic link is shown in Figure 3.18.

The optical transmitter consists of an LED coupled directly to the optical fibre. The LED is supplied with pulses of current from a computer interface. The pulses of current produce pulses of light that travel along the fibre until they reach the optical receiver unit.

The optical receiver unit consists of a photodiode (or phototransistor) that passes a relatively large current when illuminated and hardly any current when not. The pulses of current at the transmitting end are thus replicated at the receiving end.

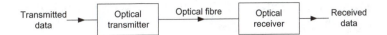

Figure 3.18 *A one-way (unidirectional) optical fibre data link*

A bidirectional optical fibre link can be produced by adding a second fibre and having a transmitter and receiver at each end of the link (Figure 3.19). A simple optical fibre data link of this sort is capable of operating at data rates of up to 500 Kbytes/s and distance of up to 1 km. More sophisticated equipment (using high-quality low-loss fibres) can work at data rates of up to 250 Mbytes/s and at distances of more than 10 km.

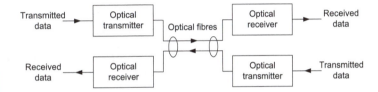

Figure 3.19 *A two-way (bidirectional) optical fibre data link*

Activity 3.13

There are several different types of optical fibre. Identify at least three different types and explain why they are different. Present your findings in the form of a series of word-processed data sheets.

Activity 3.14

Optical fibres can provide a bandwidth that is much greater than that which can be obtained using conventional copper cables. Write a brief word-processed report explaining why this is.

Case study   **Polymers**

Polymers are large molecules made of small, repeating molecular building blocks called *monomers*. The term polymer is a composite of the Greek words *poly* and *meros*, meaning 'many parts'. The process by which monomers link together to form a molecule of a relatively high molecular mass is known as *polymerisation*.

Polymers are found in many natural materials and in living organisms. For example, proteins are polymers of amino acids, cellulose is a polymer of sugar molecules and nucleic acids, such as deoxyribonucleic acid (DNA), are polymers of nucleotides. Many man-made materials, including nylon, paper, plastics and rubbers, are also polymers. Plastics are often divided into two main groups: *thermoplastics* and *thermosetting plastics* (or *thermosets*).

There is an important difference between these two classes of material. Because thermosetting plastics undergo a chemical change during moulding, they cannot be softened by reheating. Thermoplastics, on the other hand, become soft whenever they are reheated. Thermoplastics are softer and more pliable than thermosetting plastics which tend to be hard and brittle.

The chemical change experienced by thermosetting plastic materials (referred to as *curing*) is brought about from the temperature and pressure that is applied during the moulding process.

The properties of some common thermoplastics are shown in Table 3.2 whilst the properties of some common thermosetting polymers are given in Table 3.3.

Material	Relative density (N/mm^2)	Tensile strength (J)	Elongation (%)	Impact strength	Maximum service temperature (°C)
Polyamide (nylon)	1.14	50–85	60–300	1.5–15	120
Polythene	0.9	30–35	50–600	1–10	150
Polypropylene	1.07	28–53	1–35	0.25–2.5	65–85
Polystyrene	1.4	49	10–130	1.5–18	70
PVC	1.3	7–25	240–380	–	60–105
Teflon (PTFE)	2.17	17–25	200–600	3–5	260

Table 3.2 *Properties of some common thermoplastics*

Material	Relative density (N/mm^2)	Tensile strength (J)	Elongation (%)	Impact strength	Maximum service temperature (°C)
Epoxide	1.15	35–80	5–10	0.5–15	200
Urea formaldehyde	1.50	5–75	1.0	0.3–0.5	75
Polyester	2.00	20–30	0	0.25	150
Silicone	1.88	35–45	30–40	0.4	450

Table 3.3 *Properties of some common thermosetting polymers*

Test your knowledge 3.15

Explain the difference between thermoplastics and thermosets.

Activity 3.15

A company called DuPont invented Mylar polyester film in the early 1950s. Investigate the development and use of Mylar in the manufacture of electronic components. Write a brief word-processed report giving your findings.

Case study

Composite materials

In recent years, composite materials have had a huge impact on the automotive, marine, aerospace and construction sectors. It is important to realise, at the outset, that composite materials are quite different from metals. As their name suggests, composite materials are combinations of materials that differ in composition and form where the individual materials retain their separate identities and do not dissolve or merge together.

In a composite material, the separate constituents act together to provide the necessary mechanical strength or stiffness to the composite part. A modern example of a composite material is *glass-reinforced plastic* (GRP) where the glass fibres reinforce the epoxy resin matrix. On its own, the epoxy resin is a relatively weak, brittle material, and although glass fibres are strong and stiff, they can only be loaded in tension as a bare fibre. When combined, the epoxy resin and fibre give a strong, stiff composite material with excellent toughness and relatively little weight.

Composite materials are not a new idea. Mud huts were constructed from a mixture of mud, straw, grass and sticks. The lath and plaster walls of old houses were also made from a form of composite. There are also naturally occurring composite materials such as wood. Today, carbon, aramid and glass fibres have replaced grass and straw whilst epoxy resins have replaced mud and cow dung!

Reinforced concrete is another good example of a composite material. Both the steel and concrete materials retain their original identity in the finished structure. However, because they work together, the steel carries the tension loads and the concrete carries the compressive load. Reinforced concrete is very widely used in the construction industry for motorways, bridges and other large structures.

The mechanical properties of composite materials are superior to and sometimes uniquely different from the properties of their constituents (Figure 3.20). In the case of GRP used in the aerospace and automotive sectors the advantages of composite materials (when compared with sheet metal) can be summarised as follows:

- complex structures and shapes can be easily fabricated (important for both aesthetic and aerodynamic reasons)

Key point

GRP and steel-reinforced concrete are both examples of composite materials.

Key point

Composite materials consist of two or more physically distinct and mechanically separable materials manufactured in such a way as to ensure controlled and even dispersion of the constituents.

Key point

The mechanical properties of composite materials are superior to those of their constituents. For example, when combined in GRP, epoxy resin and glass fibre produce a material which is strong, stiff and tough.

Figure 3.20 *Composite materials are widely used in the aerospace industry to produce lightweight aerodynamic profiles*

- ability to integrate several parts into a single large component (impossible to do this with metal construction)
- freedom from corrosion
- combination of lightness and strength.

The ability to integrate parts together (known as *parts integration*) can lead to several further advantages including a reduced parts count, lower costs and faster assembly time. By virtue of their higher performance for a given weight, composites also offer advantages in terms of fuel savings. This is a particularly important consideration in the aerospace sector. The ability to produce very smooth aerodynamic profiles is another important consideration when aircraft or road vehicles are designed to operate economically or at high speed.

Despite these considerable advantages, it is important to mention also that composites have some disadvantages. These include the fact that composites can often be brittle and that repair of composite parts and structures requires specialised techniques and may take some considerable time whilst hot or cold *curing* takes place.

Activity 3.17

Visit Web sites devoted to Thrust SSC, a high-speed land vehicle. Determine how composite materials have contributed to the exceptional performance and, in particular, the drag resistance, of this vehicle and present your findings in the form of a brief written report. A useful interview with Thrust's designer, Ron Ayers, can be found at http://ourworld.compuserve.com/homepages/andy-graves/ronayers.htm

Case study

The microprocessor

The information revolution has largely been made possible by developments in electronics and in the manufacture of integrated circuits in particular. Ongoing improvements in manufacturing technology have given us increasingly powerful integrated circuit chips. Of these, the microprocessor (a chip that performs all of the essential functions of a computer) has been the most notable development.

A microprocessor is a single chip of silicon that performs all of the essential functions of a computer *central processing unit* (CPU) on a single silicon chip. Microprocessors are found in a huge variety of applications including engine management systems, environmental control systems, domestic appliances, video games, fax machines, photocopiers, etc.

The CPU performs three functions: it controls the system's operation; it performs algebraic and logical operations; and it stores information (or *data*) whilst it is processing. The CPU works in conjunction with other chips, notably those that provide random access memory (RAM), read-only memory (ROM) and input/output (I/O).

The key process in the development of increasingly powerful microprocessor chips is known as *microlithography*. In this process the circuits are designed and laid out using a computer before being photographically reduced to a size where individual circuit lines are about 1/100 the size of a human hair. Early miniaturisation techniques, which were referred to as large-scale integration (LSI), resulted in the production of the first generation of 256 K-bit memory chip (note that such a chip actually has a storage capacity of 262,144-bits where each bit is a binary 0 or 1). Today, as a result of very-large-scale integration (VLSI – Figure 3.21), chips can be made that contain more than a million transistors.

One of the most popular microprocessors to appear in the last 20 years was originally conceived in 1977 by a project team at Motorola. The principal architect of this chip, the 68,000, was a man called Tom Gunter. The project was known as 'Motorola Advanced Computer System on Silicon' (MACSS). At the time, Motorola was considering what direction to take in the development of their existing 8-bit

Key point

The microprocessor is an example of how today's integrated circuit technology has made it possible to fabricate many millions of electronic components on a single silicon chip.

Test your knowledge 3.18

What do each of the following abbreviations stand for?

(a) CPU
(b) RAM
(c) ROM
(d) LSI
(d) VLSI.

Figure 3.21 *A VLSI chip*

Activity 3.18

The die used to produce a microprocessor is cut from a thin wafer of silicon on which a number of identical chips are fabricated. Draw a series of sketches that illustrate the process of chip manufacture, starting with a cylinder of pure silicon and ending with a chip encapsulated in a package with pins that allow it to be connected to a socket mounted on a printed circuit board.

microprocessors. Tom Gunter proposed a 16-bit microprocessor extendible to a full 32 bits. At the time, this was a radical departure from the current state of the art, which centred around 8-bit microprocessors employed in systems with a somewhat limited memory capacity (64 Kbytes maximum).

Tom Gunter proved to be a visionary with his proposal for a highly complex microprocessor (containing the equivalent of approximately 68,000 transistors). Gunter's device could only be manufactured in 3 micron (μm) HMOS (a process which had not, at the time, been perfected). It would employ 32-bit data paths and two separate arithmetic logic units (ALUs). Furthermore, the microprocessor would require a massive 64-pin package (only 40-pin packages had been used at that time) and the die used in the production of the semiconductor chip would have to be very much larger than anything that had ever been used before.

It was something of a testimonial to Motorola's faith in its development team (which included development, software and fabrication engineers) that the project to develop the 68000 went ahead. The concept was to create a new family of super-microprocessors that would provide increasing levels of functionality and performance.

What material is used in the manufacture of integrated circuits?

At the time, it was felt that the majority of future software development would be in high-level languages and that this should be reflected in both the internal architecture and instruction set of the 68,000. Thus, the 68,000 chip was given a particularly sleek and uncluttered architecture.

The 68,000 employs what has become commonly known as a *complex instruction set*. Computers that employ this technology are known as CISC (complex instruction set computer) machines.

Activity 3.19

Find out about the HMOS process. Explain what it is and how it is used in the production of microprocessors and other VLSI chips. Present your findings in the form of a brief technical report.

Activity 3.20

CISC machines are not the only types of computer. Find out about the alternative RISC technology. Why is it different and how does it compare? Present your findings in the form of a brief article for a local computer club.

3.3.4   Automation and control systems

Automation has been widely introduced into the engineering industry and can be defined as the use of mechanical and electrical/electronic systems to carry out processes and functions without requiring direct human control. Automation is often associated with the use of robots but it can also be applied to manufacturing processes that operate under computer (programmed) control. Our first case study introduces a particular example of automation in the development of remotely operated vehicles (ROVs).

Case study

Remotely operated vehicles

ROVs are designed to work in environments in which humans could not survive. Such environments include the depths of the world's oceans as well as outer space. If you saw a film of the wreck of the *Titanic* or the recent landing on Mars you will already have experienced the use of the vehicles!

There is no operator or driver actually sitting in an ROV, instead a remote control system is used to operate the vehicle and the 'pilot' remains at some distance from the point at which the vehicle is operating (a very considerable distance in the case of the Mars lander).

Many different companies are involved with the design, manufacture and operation of ROVs. A typical ROV system has at least four major system components. They are:

- the vehicle chassis and drive mechanism
- a deployment system that gets the vehicle to where it is to be used (and also recovers it where possible)
- a remote control system which allows the operator to work remotely from the vehicle
- a cable, fibre or radio telemetry system that links the vehicle with the control console.

ROVs are able to perform many tasks including:

- marine surveying, search and recovery
- deep-sea cable burial, inspection, maintenance and repair
- marine engineering, construction and repair
- space and interplanetary exploration
- fire fighting (particularly in the chemical and oil industries)
- inspection, maintenance and repair in the nuclear industry
- bomb disposal.

ROV systems usually require advanced materials and construction methods, and are usually 'one-of-a-kind' vehicles. For example, many deep-sea ROVs are designed for operation at a particular depth. ROVs for space exploration, on the other hand, are designed to operate under an extreme range of temperatures and are usually designed for a single mission (they may often not be recoverable).

The typical specification for a deep-sea ROV designed for surveying, search and recovery is shown in Table 3.4.

Test your knowledge 3.20

State THREE applications of ROVs.

Activity 3.21

Use the Internet or other information sources to locate information and specifications of at least three deep-sea ROVs. Compare these specifications and provide details of typical projects in which these vehicles have been used. Present your findings in the form of a series of word-processed fact sheets.

Depth	2 km
Lifting capacity	670 kg (without additional buoyancy)
Payload	225 kg (without additional buoyancy)
Propulsion	Eight hydraulic thrusters (two axial, four vertical and two lateral)
Forward speed	3 kn (1.5 m/s) at 1000 lbs (4.5 kN) thrust
Hydraulic power	Twin 50 HP (37.5 kW) hydraulic power units
Cameras	Colour camera with pan and tilt plus wide angle black and white camera
Instruments	Sonar, depth and gyro heading sensors
Illumination	Six 250 W lights
Manipulators	Two 7-function manipulators
Cutters	Abrasive wheel (capable of cutting up to 76 mm diameter cable) plus hydraulic guillotine

Table 3.4 *Typical specification for a deep-sea ROV*

Activity 3.22

Use the Internet or other electronic information sources to find out how a gyro works. Present your findings in the form of a brief illustrated article for an engineering club newsletter.

Activity 3.23

Use your school or college library or other paper-based information sources to find out how sonar works. Present your findings in the form of a brief illustrated article for an engineering club newsletter.

Activity 3.24

Sketch a labelled diagram showing the propulsion system used in the ROV described in Table 3.4. The lifting capacity of the ROV is to be increased. Suggest, with the aid of a diagram, a method of providing additional buoyancy. Present your work in the form of a series of overhead projector transparencies.

Test your knowledge 3.21

List the four main system components of an ROV. Explain briefly what each system component is used for.

One of the payloads carried by the recent Mars Pathfinder spacecraft was a miniature ROV called 'Sojourner'. The ROV was deployed on the surface of Mars on 4th July 1997 following a 7-month journey through interplanetary space. After landing the microrover carried out a variety of experiments to determine wheel–soil interactions, navigation and hazard avoidance capabilities. Sojourner also carried an alpha proton X-ray spectrometer (APXS) that was used to determine the element composition of the soil and rocks on Mars.

Sojourner was a six-wheeled vehicle of a rocker bogie design (see Figure 3.22). The ROV weighed a mere 11.5 kg (25 lbs) and was about the size of a milk crate. The six-wheeled design allowed the ROV to traverse obstacles of up to a wheel diameter (13 cm) in size. Each wheel was independently actuated and geared (2000:1) providing superior climbing capability in soft sand. The front and rear wheels were independently steerable, providing the capability for the vehicle to turn in place. The vehicle had a top speed of 0.4 m/min.

Figure 3.22 *Sojourner*

Sojourner was powered by a 0.22 m² solar panel comprising of 13 sets of 18, 5.5 mm Gallium Arsenide (GaAs) solar cells (see Table 3.5). The solar panel was backed up by nine Lithium–Chloride primary batteries, providing up to 150 Whr. This combined panel/batteries system allowed Sojourner's system to demand up to 30 W of peak power. The normal driving power requirement for the microrover was a modest 10 W.

Since the temperature on the surface of Mars can fall to −110°C during the Martian night, some of Sojourner's components were enclosed in a warm 'electronics box' (Web). The Web was insulated, coated with high and low emissivity paints, and heated with a combination of three 1 W resistive heater units and waste heat produced by the other electronics components. The design allowed the Web to keep

Construction	Gallium arsenide on germanium (GaAs/Ge)
Configuration	13 parallel strings, each with 18 series-connected cells
Operating voltage	14–18 V
Temperature	−140°C to +110°C (max)
Power output	16.5 W on Mars (at noon)
Hydraulic power	Twin 50 HP (37.5 kW) hydraulic power units
Substrate	Nomex honeycomb
Weight	0.34 kg
Area	0.22 m^2
Coverglass	3 mm
Efficiency	>18%

Table 3.5 *Specification for Sojourner's solar cell battery*

Sojourner's electronic components at an ambient temperature of between −40°C and +40°C during a complete Martian day.

Control was provided by an 80C85 microprocessor which was capable of executing 100,000 instructions per second. The microprocessor system was fitted with 176 Kbytes of PROM and 576 Kbytes of RAM and 70 I/O channels for sensors and devices such as cameras, modem, motors and experimental electronics.

Vehicle motion control was accomplished through the on/off switching of the drive or steering motors. An average of motor encoder (drive) or potentiometer (steering) readings determined when to switch off the motors. When the motors were off, the microprocessor system was programmed to carry out a proximity and hazard detection function, using its laser and camera system to determine the presence of obstacles in its path. The vehicle was able to avoid obstacles in its path without operator intervention and yet still arrive at its commanded goal location. While stopped, the microprocessor system updated its measurement of distance travelled and heading using the average speed and the on-board gyro. This allowed the microrover to make regular estimates of its progress to the goal location.

Command and telemetry were provided by modulator–demodulators (modems) on the microrover and lander. The microrover was linked to the lander by means of a short-range UHF radio system. During the Martian day, the microrover regularly requested transmission of any commands sent from earth that were stored in the lander's memory. When commands were not available, the microrover transmitted any telemetry collected during the last interval between communication sessions. The telemetry received by the lander from the microrover was stored and forwarded to the earth as part of the lander's own telemetry. In addition, this communication system was used to provide a 'heartbeat' signal during vehicle driving.

When stopped, the microrover sent signals to the lander. Once acknowledged by the lander, the microrover then moved to the next

Key point

ROVs can go into places that are inaccessible or unsafe for humans.

Test your knowledge 3.22

Explain how a typical ROV is able to accept commands from a remote location.

stopping point along its track. Commands for the microrover were sent, via the lander, from the mission's earth control station. In the opposite direction, the lander transmitted its stereo images of the microrover back to the earth. These images, portions of a terrain panorama and supporting images from the microrover's own cameras were displayed at the control station. The operator was then able to designate the display points in the terrain that were to serve as goal locations for the microrover. The coordinates of these points were transferred into a file containing the commands for execution by the microrover and then sent from earth to the lander for onward transmission to the microrover.

In addition, the operator was able to use a model which, when overlayed on the image of the vehicle, measured the location and heading of the ROV. This information was also transferred into the command file to be sent to the microrover on its next traverse in order to correct any navigation errors.

Activity 3.25

Use the Internet or other electronic information sources to find out how a modem works. Present your findings in the form of a brief illustrated article for an engineering club newsletter.

Activity 3.26

Use your school or college library, electronic data books, or other paper-based information sources to obtain data on the 80C85 microprocessor (including, as a minimum, absolute maximum ratings and pin connections). Present your findings in the form of an illustrated data sheet.

Activity 3.27

Sketch a labelled diagram showing the communication systems used to:

(a) send commands from the earth control station to Sojourner
(b) receive telemetry data sent from Sojourner at the earth control station.

Present your work in the form of a drawing produced by a CAD or computer drawing package.

Case study

Microprocessor systems

Earlier, when we looked at new material and components, we introduced the microprocessor. In our last case study, we mentioned the microprocessor system that was used to control the Sojourner ROV. These sophisticated VLSI chips are now found in a huge variety of applications, from controlling the engines on a large passenger jet to operating a CNC machine tool. However, even the most sophisticated microprocessor chip is of little use unless it is supported by a number of other devices connected in what we refer to as a *microprocessor system*.

The first microprocessor systems were developed in the early 1970s. These were simple and crude by today's standards but they found an immediate application in the automotive industry where they were deployed in engine management and automatic braking systems. Today, microprocessor systems are found in a huge variety of applications from personal computers to washing machines!

Figure 3.23 shows the block schematic diagram of a complete microprocessor system. All such systems, regardless of their complexity, conform to this basic arrangement.

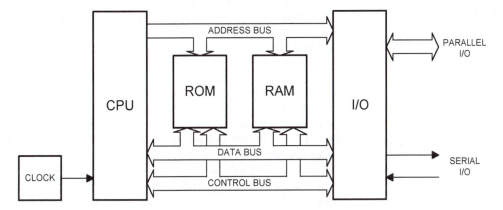

Figure 3.23 *A block schematic diagram of a microprocessor system*

The CPU is generally the microprocessor chip itself. This device contains the following main units:

- storage locations (called *registers*) that can be used to hold instructions, data and addresses during processing
- an *arithmetic logic unit* that is able to perform a variety of arithmetic and logical function (such as comparing two numbers)
- a *control unit* which accepts and generates external control signals (such as *read* and *write*) and provides timing signals for the entire system.

In order to ensure that all the data flow within the system is orderly, it is necessary to synchronise all of the data transfers using a clock signal. This signal is often generated by a clock circuit (similar to the clock in a digital watch but much faster). To ensure accuracy and stability the clock circuit is usually based on a miniature quartz crystal.

Figure 3.24 *This microprocessor system (based on an 8-bit microprocessor) is used to control a computer printer*

All microprocessors require access to read/write memory in which data (e.g. the results of calculations) can be temporarily stored during processing. Whilst some microprocessors (often referred to as microcontrollers) contain their own small read/write memory, this is usually provided by means of a semiconductor *random access memory* (RAM).

Microprocessors generally also require more permanent storage for their *control programs* and, where appropriate, operating systems and high-level language interpreters (Figure 3.24). This is usually provided by means of semiconductor *read-only memory* (ROM).

To fulfil any useful function, a microprocessor system needs to have links with the outside world. These are usually supplied by means of one, or more, VLSI devices which may be configured under software control and are therefore said to be *programmable*. The *input/output* (I/O) devices fall into two general categories: *parallel* (where a *byte* is transferred at a time along eight wires), or *serial* (where one *bit* is transferred after another along a single wire).

The basic components of a microprocessor system (CPU, RAM, ROM and I/O) are linked together using a multiple connecting arrangement known as a *bus*. The *address bus* is used to specify memory locations (i.e. addresses), the *data bus* is used to transfer data between devices, and the *control bus* is used to provide timing and control signals (such as *read* and *write*, *reset* and *interrupt*) throughout the system.

Case study

Programmable logic controllers

PLCs are microprocessor systems that are used for controlling a wide variety of automatic processes, from operating an airport baggage handling system to brewing a pint of your favourite lager! PLCs are rugged and modular and they are designed specifically for operating in a process control environment.

The control program for a PLC is usually stored in one or more semiconductor memory devices. The program can be entered (or modified) using a simple hand-held programmer, a laptop computer, or downloaded from a local area network (LAN). PLC manufacturers include Allen Bradley, Siemens and Mitsubishi.

A typical PLC is shown in Figure 3.25. Input/output (I/O) connections are made via the terminal strips on each edge of the unit. A bank of LED status indicators are used to display the state of each of the I/O lines (i.e. whether they are *on* or *off*) whilst a round multi-pin DIN connector is used to connect the hand-held programmer (see Figure 3.26).

Key point

PLCs provide a rugged and flexible means of controlling a wide variety of industrial processes.

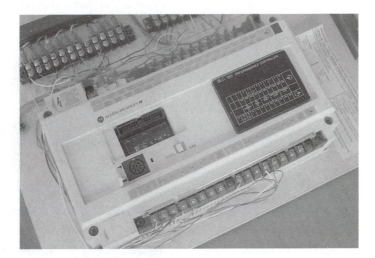

Figure 3.25 *A typical PLC*

In operation, the PLC's control program repeatedly scans the state of each of its inputs (to determine whether they are *on* or *off*) before using this information to decide on what should happen to the state of each of its output lines.

The PLC usually derives its inputs from switches (such as keypads, buttons, or contacts that make and break) but it can also derive its input from sensors (e.g. liquid level sensors, temperature sensors, position sensors, motion sensors, etc.).

Many PLCs can also be fitted with modules that allow analogue signals (i.e. signals that can vary continuously in voltage level rather than just be *on* and *off*) to be used both for input and for output. In the former (input) case an *analogue-to-digital converter* (ADC) is required whilst in the latter (output) case a *digital-to-analogue converter* (DAC) is needed.

Figure 3.26 *A typical hand-held PLC programmer*

A typical PLC application in which a PLC is controlling a small conveyor belt is shown in Figure 3.27. Note the start and stop buttons and the optical position sensors mounted at the side of the conveyor belt.

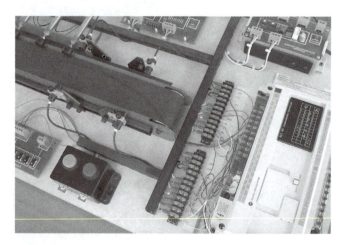

Figure 3.27 *A typical PLC system in which the PLC controls a small conveyor belt*

Test your knowledge 3.25

Give TWO applications for PLCs.

Test your knowledge 3.26

Explain, in simple terms, how a PLC operates.

Activity 3.29

Investigate a PLC system in your school or college (if your school does not have such a system they may be able to arrange for you to visit a local engineering manufacturer or a nearby Further Education college that has this equipment). Find out what the PLC is used for and how it is programmed. Write a brief report describing what you have seen and explain, in simple terms, how a PLC works. Illustrate your report with relevant sketches, diagrams and photographs.

3.4 Stages in engineering and manufacturing a product

By now, you should realise that a number of stages and activities go into making an engineered product. These stages can be generally grouped as:

- design
- marketing
- production planning
- purchasing, material supply and control
- production/processing
- assembly and finishing
- packaging and dispatch.

Engineers are involved with most of these stages and particularly with the design, production planning, production/processing, and assembly and finishing stages.

When you investigate an engineered product you need to identify the main stages and activities in designing and manufacturing the product. For example, the modern radio receiver that we looked at earlier requires separate production of the plastic case parts and printed circuit board.

The printed circuit boards will need to be manufactured (cut, etched and drilled) before the components can be fitted and soldered into place. The gears will need to be fitted to the tuning drive and the loudspeaker fitted to the inside of one-half of the case.

Next, the two printed circuits will need to be attached to the mounting pillars on one-half of the case and the remainder of the wiring completed. The radio can now be tested and adjusted. Finally, the other case half is fitted and the remaining knobs are pushed onto the controls prior to packaging and dispatch to the retailer.

Activity 3.30

Examine a typical foot pump that can be used for inflating the tyres of a car. Identify each of the component parts and the materials used. Suggest the stages used in engineering and manufacturing this product and present your findings in the form of a brief written report.

3.5 Investigating products

Our final case study takes the form of a product investigation. We are going to look at a product that you are probably already very familiar with, a compact disc (CD). The technology that has made this product possible has only been available in the last 12 years.

When you investigate a product you need to be able to research information from manufacturers and suppliers, handle and examine the product, carry out a simple assessment of properties (including structure, heaviness, colour and surface finish), and evaluate the need for the technology, materials and components used. You will go on to extend this case study by carrying out your own investigation of a portable CD player.

Case study

Compact discs

CDs can provide 65-min of high-quality recorded audio or 650 Mbytes of computer data that is roughly equivalent to 250,000 pages of A4 text. It is therefore hardly surprising that the compact disc has now become firmly established in both the computing and hi-fi sectors.

As with most 'new' products, the technology used in compact discs (and in compact disc players and recorders) relies on several other technologies working together. In the case of compact discs, the most important of these are:

- digital audio technology (being able to represent audio signals using a sequence of digital codes)
- optical technology (being able to produce a precisely focused beam from a laser light source)
- control system technology (being able to control motor speed as well as the position and focus of the optical unit)
- manufacturing technology (being able to reliably and cost-effectively manufacture both the equipment used to play compact discs and the compact disc media itself).

Key point

The CD is a good example of how advances in technology have had an impact on both the product itself and the manufacturing processes that are used in its production.

Before explaining how compact disc technology works, it is worth looking at the previous storage technology in which vinyl discs were used for recording analogue signals in the form of 'long-play' (LP) and 'extended-play' (EP) 'records'. Despite its limitations, vinyl disc recording technology survived for nearly four decades (from around 1950 to around 1990). However, the problems associated with recording analogue signal variations in a groove pressed into the relatively soft surface of plastic eventually led to the downfall of the LP recording and its replacement by the digital compact disc in the late 1980s (Figure 3.28).

A comparison between the performance specifications of a compact disc system with those of an LP record player reveals a number of important differences as shown in Table 3.6.

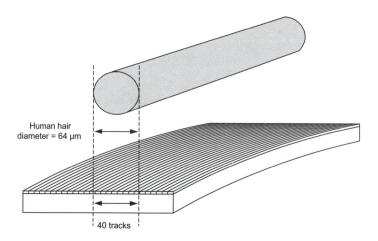

Human hair diameter = 64 μm

40 tracks

Figure 3.28 *A human hair compared with the tracks on a CD*

Specification	CD player	LP record player
Recording technique	Digital	Analogue
Material	Glass	Vinyl
Dynamic range	90 dB typical	70 dB typical
Frequency response	20 Hz to 20 kHz $\pm$ 0.5 dB	20 Hz to 20 kHz $\pm$ 3 dB
Signal-to-noise ratio	90 dB typical	60 dB typical
Harmonic distortion	Less than 0.01%	Less than 2%
Channel separation	More than 90 dB	Less than 40 dB
Wow and flutter	Better than 1 part in 105	Better than 0.1%
Diameter	120 mm	305 mm
Rotational velocity	196–568 rpm	33.3 rpm
Playing time (per side)	65 min typical	20 min typical

Table 3.6 *Comparison of CD and LP record player specifications*

Conventional CD-ROMs, like audio compact discs, are made up of three basic layers. The main part of the disc consists of an injection-moulded polycarbonate *substrate* which incorporates a spiral track of *pits* and *lands* that are used to encode the data that is stored on the disc (Figure 3.29). Over the substrate is a thin aluminium (or gold) reflective layer, which in turn is covered by an outer protective lacquer coating.

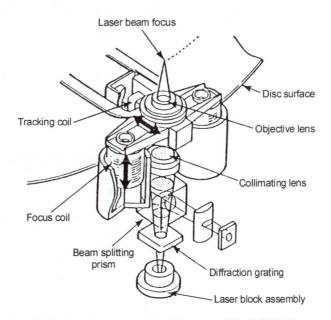

Figure 3.29 *Optical assembly fitted to a CD-ROM drive*

Information is retrieved from the CD by focusing a low-power (0.5 mW) infrared (780 nm) laser beam onto the spiral track of pits and lands in the disc's substrate. The height difference between the pits and the adjacent lands creates a phase shift causing destructive interference in the reflected beam. The effect of the destructive interference

Test your knowledge 3.28

Give THREE reasons why compact discs offer vastly improved performance when compared with LP records.

Test your knowledge 3.29

What do the following abbreviations stand for?

(a) CD
(b) CD-ROM
(c) CD-R
(d) CD-E.

Test your knowledge 3.30

Explain why precise focusing of the laser beam in a CD player is essential.

and light scattering is that the intensity of the light returned to a photodiode detector is modulated by the digital data stored on the disc. This modulated signal is then processed, used for tracking, focus and speed control, and then decoded and translated into usable data. The optical assembly, complete with lenses and focus coils, is shown in Figure 3.30.

Conventional CDs and CD-ROMs only support playback (or reading) of the data stored in them. In recent years new technology has appeared which supports both playback (reading) and recording (writing). This technology has resulted in two new types of compact disc: the CD-R (recordable) and the CD-E (erasable).

Figure 3.30 *CD player optical unit showing objective lens (centre), focus coil (above centre) and tracking coil (right of centre)*

Activity 3.31

Investigate the construction of a typical portable CD player. Itemise each of the main components or sub-assemblies used and describe the materials and processes used in its manufacture. Give at least THREE examples of how modern technology has had an impact on the design and manufacture of the CD player and comment on the technology and processes that it has replaced. Present your findings in the form of a brief class presentation using appropriate visual aids and handouts incorporating drawings, diagrams and sketches.

Review questions

1. For each of the engineering sectors listed below, identify a product or service which is based on the application of new technology and, in each case, explain briefly how the technology in question has

influenced the development of the product:

(a) Information and communications technology
(b) New materials and components
(c) Automation.

2. Explain the following terms used in computer aided engineering (CAE):

(a) Computer aided design (CAD)
(b) Computer aided manufacture (CAM)
(c) Computer integrated manufacturing (CIM)
(d) Computer numerical control (CNC).

3. Explain what a database is and how it is used in a typical engineering company.

4. Explain the following terms in relation to the Internet and the World Wide Web:

(a) Uniform resource locator (URL)
(b) Hypertext transfer protocol (HTTP)
(c) Hypertext markup language (HTML).

5. Explain how an engineering company can use an Intranet to improve communications between employees and departments.

6. With reference to the Internet and the World Wide Web, explain the function of a search engine.

7. Explain how a file can be sent using e-mail.

8. With the aid of a diagram, explain how data can be transferred from one computer to another using an optical fibre link.

9. Explain the essential difference between thermoplastics and thermosetting materials (thermosets).

10. Give the names of TWO thermoplastic materials and TWO thermosetting materials.

11. Categorise each of the following materials as either a thermoplastic material or a thermosetting material:

(a) Polythene
(b) Polyester
(c) Silicone
(d) Teflon.

12. State a typical application for each of the materials in Question 11.

13. Explain what is meant by a composite material.

14. Name TWO composite materials used in engineering and give a typical application for each material.

15. State THREE advantages of using composite material in the body shell of a high-performance road vehicle.

16. Describe THREE applications of ROVs.

17. Explain how data is stored on the surface of a compact disc.

18. Identify THREE different uses of new technology within a compact disc player.

19. Explain how GPS technology can be used to provide accurate location information for any point on the earth's surface.

20. Figure 3.31 shows an internal view of a power supply for use in a PC. Describe THREE ways in which technology is used in the manufacture of this unit.

Figure 3.31 *See Question 20*

21. Explain the term 'engineering fabrication'. Give TWO examples of products that come from this engineering sector.

22. In relation to a database, explain each of the following terms:

(a) Record
(b) Field
(c) Key.

23. Describe an application for a database in a typical engineering company.

24. Give TWO examples of products that are associated with:

(a) The mechanical engineering sector
(b) The automotive sector
(c) The electrical/electronic engineering sector.

25. Figure 3.32 shows the internal construction of a modern radio receiver. Identify materials used for:

(a) The external case
(b) The gears used for the tuning drive
(c) The integrated circuit
(d) The loudspeaker frame
(e) The printed circuit board.

Figure 3.32 *See Question 25*

26. Describe, with the aid of an example, how control technology is used in an engineering manufacturing process.

27. Relate each engineering product in List A to the manufacturing sector in List B:

List A (products)	List B (sectors)
GPS receivers	Aerospace
Gearboxes	Telecommunications
Gas turbine blades	Automotive
Metal ducting	Chemical engineering
Fertiliser	Mechanical fabrication
Microwave relays	Electrical/electronic

Chapter 4　Maths and science for engineering

Summary

Maths and science underpins all aspects of engineering. Due to this, you should consider maths and science to be essential 'tools' that will allow you to understand and progress within engineering. This topic-based chapter provides you with the necessary maths and science underpinning knowledge in order to complement your GCSE in Engineering course. Whilst maths and science are not part of the GCSE course (you will almost certainly be studying both of these subjects separately) they become really important if you intend to study engineering at a higher level.

The examples in this chapter all relate to engineering applications and they are designed to help you to understand the concepts and underlying scientific principles introduced elsewhere in the book. Typical examples might include determining the volume of liquid in a storage tank, estimating the force on a structural member, or the acceleration of a motor vehicle on a test track. The topics covered include:

- Numbers, decimals and fractions
- Units, multiples and sub-multiples
- Engineering notation, standard form
- Solution of engineering formulae
- Measurement of area, length and volume
- Force, mass and acceleration
- Gravity and weight
- Angular measure
- Trigonometry
- Graphs.

4.1 Numbers

Let us begin with numbers. Engineers tend to use a lot of numbers for the simple reason that they are *precise*. For example, instead of saying that we need a 'large storage tank' we would convey a lot more meaning (and be much more *specific*) if we say that we need 'a tank with a capacity of 4500 litres'. In fact, when engineers draw up a *specification* they do so using numbers and drawings in preference to written descriptions.

Whole numbers that have a positive (+) sign, such as 1, 2, 3, 4, …, are known as *positive integers*. Negative numbers that have a minus (−) sign, such as −1, −2, −3, −4, …, are known as *negative integers*. Note that if a number is positive, we do not usually include a positive (+) sign to show that it is positive. Instead, we simply assume that it is there!

The number of units that a number is from zero (regardless of its direction or sign) is known as its *absolute value*. Positive numbers are conventionally shown to the right of the *number line* (see Figure 4.1) whilst negative numbers are shown to the left. When the sign is shown these numbers are said to be *signed*. For example, the number +5 has an absolute value of 5 and its sign is + (positive). Similarly, the number −3 has an absolute value of 3 and its sign is − (negative). Note that the number zero (0) is unique in that it is neither a positive integer nor is it a negative integer.

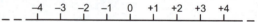

Figure 4.1 *The number line (showing positive and negative integers)*

When we are performing arithmetic using numbers we do need to be careful when we show the signs. For example, if we need to find the sum of the first three positive integers (1, 2 and 3) we would write this as follows:

Sum of first three positive integers = 1 + 2 + 3 = 6

However, if we are asked to find the sum of the first three negative integers (−1, −2 and −3) we would write:

Sum of first three negative integers = (−1) + (−2) + (−3)
= −1 − 2 − 3 = −6

Notice how we have used brackets to help to clarify the arithmetic.

As we frequently have to deal with numbers that lie between two integer numbers, integers are often not precise enough for use in engineering applications. We can get over this problem in two ways; using *fractions* and using a *decimal point*. For example, the number that sits mid-way between the positive integers 3 and 4 can be expressed as 3½ or 3.5. Similarly, the number that sits equally between −1 and −2 can be expressed as −1½ or −1.5 (see Figure 4.2). Table 4.1 details some common fractions and their corresponding decimal value.

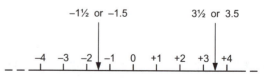

Figure 4.2 *Decimal and fractional numbers shown on the number line*

Fraction	Decimal value
$\dfrac{1}{2}$	0.5
$\dfrac{1}{4}$	0.25
$\dfrac{1}{8}$	0.125
$\dfrac{1}{10}$	0.1
$\dfrac{1}{16}$	0.0625
$\dfrac{1}{100}$	0.01
$\dfrac{1}{1000}$	0.001

Table 4.1 *Common fractions with decimal equivalents*

Test your knowledge 4.1

Find the value of each of the following expressions:

(a) $2 - 13$
(b) $-7 + 19$
(c) 5×-9
(d) -11×-7
(e) $15 \div -3$

Test your knowledge 4.2

Which of the following statements is correct?

(a) $1 + (-1) = 0$
(b) $1 - (-1) = 0$
(c) $-1 \times +1 = 0$
(d) $-1 \times -1 = 0$
(e) $-1 \div +1 = 0$

Key point

When a positive number is multiplied by a negative number (or vice versa) the result is negative but when two negative numbers are multiplied together the result is positive.

Key point

When a positive number is divided by a negative number (or vice versa) the result is negative but when the two numbers are *both* negative the result is positive.

4.1.1 Laws of signs

There are four basic laws for using signs. They are:

First law
To add two numbers with like signs, add their absolute values and prefix their common sign to the result.

   Example: $2 + 3 = (+2) + (+3) = +5 = \mathbf{5}$

Second law
To add two signed numbers with unlike signs, subtract the smaller absolute number from the larger and prefix the sign of the number with the larger absolute value to the result.

   Example: $2 + (-3) = (+2) + (-3) = -(3 - 2) = \mathbf{-1}$

Third law
To subtract one number from another, change the sign of the number to be subtracted and follow the rules for addition.

   Example: $2 - 3 = (+2) - (+3) = -(3 - 2) = \mathbf{-1}$

Fourth law
To multiply (or divide) one signed number by another, multiply (or divide) their absolute values; then, if the numbers have like signs,

prefix a plus sign to the result; if they have unlike signs, prefix a minus sign to the result.

Example: $2 \times 3 = (+2) \times (+3) = +6 = \mathbf{6}$
Example: $2 \times -3 = (+2) \times (-3) = -6 = \mathbf{-6}$
Example: $-2 \times -3 = (-2) \times (-3) = +6 = \mathbf{+6}$

4.2 Units and symbols

You will find that quite a large number of units and symbols are used in science and engineering so let us get started by introducing some of them. In fact, it is important to get to know these units and also to be able to recognise their abbreviations and symbols before you actually need to use them. Later we will explain how these units work in much greater detail, but for now we will simply list some of the most useful units and symbols so that at least you can begin to get to know something about them.

The units seen in Table 4.2 are called *fundamental units* (or *base units*) and they are part of the International System (known as 'SI') of units.

Name	Symbol	Unit	Abbreviation
Mass	M	kilogram	kg
Length	L	metre	m
Time	T	second	s
Electric current	I	ampere	A
Temperature	θ	kelvin	K
Amount of substance	N	mole	mol
Luminous intensity	J	candela	cd

Table 4.2 *SI base units*

> **Key point**
>
> All derived units are derived in terms of the seven fundamental (base) units.

> **Key point**
>
> Symbols used for electrical and other quantities are normally shown in italic font whilst units are shown in normal (non-italic) font. Thus M and L are symbols whilst J and W are units.

Other units can be derived from these seven fundamental units. These are called derived units and a selection of some of the most common is shown in Table 4.3.

4.2.1 Multiples and sub-multiples

Unfortunately, the numbers that we meet in engineering can sometimes be very large or very small. For example, the voltage present at the antenna input of a very high frequency (VHF) radio could be as little as 0.000,001 V. At the same time, the resistance present in an amplifier stage could be as high as 10,000,000 Ω!

We can make life a lot easier by using a standard range of multiples and sub-multiples. These use a *prefix* letter that adds a *multiplier* to the quoted value (see Table 4.4).

It is important to note that multiplying by 1000 is equivalent to moving the decimal point *three* places to the *right*. Dividing by 1000,

Quantity	Unit	Abbreviation	Derivation
Energy (or *work*)	Joule	J	1 J of energy (or 'work done') is used when 1 N of force moves through 1 m in the direction of the force
Force	Newton	N	Unit of force (a force of 1 N gives a mass of 1 kg an acceleration of 1 m/s^2)
Pressure	Pascal	Pa	A pressure of 1 Pa exists when a force of 1 N is exerted over an area of 1 m^2
Power	Watt	W	A power of 1 W is developed when 1 J of energy is used in 1 s
Electric charge	Coulomb	C	An electric charge of 1 C is transferred when a current of 1 A flows for 1 s
Frequency	Hertz	Hz	A wave has a frequency of 1 Hz if one complete cycle occurs in 1 s
Velocity	metre/ second	m/s (or m s^{-1})	A body travelling at a velocity of 1 m/s moves through 1 m every second in the direction of travel
Acceleration	metre/ second/ second	m/s/s (or m/s^2 or m s^{-2})	A body travelling with an acceleration of 1 m/s^2 increases its velocity by 1 m/s every second in the direction of travel

Table 4.3 *Examples of SI derived units*

Test your knowledge 4.3

1. State the SI unit for frequency.
2. State the SI unit for electric charge.
3. State the SI unit for force
4. State the abbreviation for the SI unit of acceleration.
5. State the symbol used for electric current.
6. Twenty joules of energy are used in half a second. What power does this correspond to?
7. A military aircraft increases its speed from 250 m/s to 300 m/s in 5 s. What acceleration is this?
8. An electric current of 5 A flows for half a minute. How much charge is moved?

Prefix	Abbreviation	Multiplier
tera	T	10^{12} (1,000,000,000,000)
giga	G	10^9 (1,000,000,000)
mega	M	10^6 (1,000,000)
kilo	K	10^3 (1000)
(none)	(none)	10^0 (1)
centi	c	10^{-2} (0.01)
milli	m	10^{-3} (0.001)
micro	μ	10^{-6} (0.000,001)
nano	n	10^{-9} (0.000,000,001)
pico	p	10^{-12} (0.000,000,000,001)

Table 4.4 *Multiples and sub-multiples*

on the other hand, is equivalent to moving the decimal point *three* places to the *left*. Similarly, multiplying by 1,000,000 is equivalent to moving the decimal point *six* places to the *right* whilst dividing by 1,000,000 is equivalent to moving the decimal point *six* places to the *left*.

Key point

To multiply a number by 1000 (one thousand) we move the decimal point three places to the right. To divide a number by 1000 we move the decimal point three places to the left.

Key point

To multiply a number by 1,000,000 (one million) we move the decimal point six places to the right. To divide a number by 1,000,000 we move the decimal point six places to the left.

Test your knowledge 4.4

1. An amplifier requires an input voltage of 50 mV. Express this in V.
2. An aircraft strut has a length of 1.25 m. Express this in mm.
3. A marine radar operates at a frequency of 9.74 GHz. Express this in MHz.
4. A generator produces a voltage of 440 V. Express this in kV.
5. A manufacturing process uses a coating with a thickness of 0.075 mm. Express this in μm.
6. A radio transmitter has a frequency of 15.62 MHz. Express this in kHz.
7. A current of 570 μA flows in a transistor. Express this current in mA.
8. A capacitor has a value of 0.22 μF. Express this in nF.
9. A resistor has a value of 470 kΩ. Express this in MΩ.
10. A plastic film has a thickness of 0.0254 cm. Express this in mm.

Example 4.1

A high-speed vehicle test track has a total distance of 3.75 km. Express this in metres.

To convert from km to m we need to apply a multiplier of 10^3 or 1000. Thus, to convert 3.75 km to m, we multiply 3.75 by 1000, as follows:

$$3.75\,\text{km} = 3.75 \times 1000 = \mathbf{3750\,m}$$

Example 4.2

An LED requires a current of 15 mA. Express this in A.

To convert mA to A, we apply a multiplier of 10^{-3} or 0.001. Thus, to convert 15 mA to A, we multiply 15 by 0.001, as follows:

$$15\,\text{mA} = 15 \times 0.001 = \mathbf{0.015\,A}$$

Note that multiplying by 0.001 is equivalent to moving the decimal point three places to the *left*.

Example 4.3

An insulation tester produces a voltage of 2750 V. Express this in kV.

To convert V to kV we apply a multiplier of 10^{-3} or 0.001. Thus we can convert 2750 V to kV as follows:

$$2750\,\text{V} = 2750 \times 0.001 = \mathbf{2.75\,kV}$$

Here again, multiplying by 0.001 is equivalent to moving the decimal point three places to the *left*.

Example 4.4

A capacitor has a value of 27,000 pF. Express this in μF.

There are 1,000,000 pF in 1 μF. Thus, to express the value 27,000 pF in μF we need to multiply by 0.000,001. The easiest way of doing this is simply to move the decimal point six places to the left. Hence 27,000 pF is equivalent to **0.027 μF** (note that we have had to introduce an extra zero before the 2 and after the decimal point).

4.2.2 Unit conversion

Unfortunately, the standard SI units are not used universally and several other types of unit are in common use. To overcome this problem we can use conversion factors to convert from one type of unit to another. Some common conversion factors are listed in Table 4.5.

Example 4.5

An aircraft flies at an altitude of 37,000 ft. Express this in metres.

From	To	Conversion factor (multiply by)
miles per hour (mph)	metres per second (m/s)	0.4470
metres per second (m/s)	miles per hour (mph)	2.2369
miles per hour (mph)	kilometres per hour (km/h)	1.6093
kilometres per hour (km/h)	miles per hour (mph)	0.6214
litres (l)	cubic metres (m^3)	0.001
cubic metres (m^3)	litres (l)	1000
feet (ft)	metres (m)	0.3048
metres (m)	feet (ft)	3.2808
miles (mi.)	kilometres (km)	1.6093
kilometres (km)	miles (mi.)	0.6214
yards (yd)	metres (m)	0.9144
metres (m)	yards (yd)	1.0936
kilograms (kg)	tonne (metric)	0.001
tonne (metric)	kilograms (kg)	1000
ton (UK)	tonne (metric)	1.01605
tonne (metric)	ton (UK)	0.9842
pounds per square inch (psi)	Pascal (Pa)	6894.8
pascal (Pa)	pounds per square inch (psi)	1.4504×10^{-4}
litres (l)	gallons (gal UK)	0.21997
gallons (gal UK)	litres (l)	4.5461

Table 4.5 *SI units conversion table*

To convert from feet (ft) to metres (m) we need to multiply by a conversion factor (see Table 4.5) of 0.3048:

So 37,000 ft = 37,000 × 0.3048 = **11,277.6 m**

4.3 Notation

Standard notation is used in mathematics to simplify the writing of mathematical expressions. This notation is based on the use of symbols that you will already recognise including: = (equal), + (addition), − (subtraction), × (multiplication), and ÷ (division). Other symbols that you may not be so familiar with include < (less than), > (greater than), ∝ (proportional to) and $\sqrt{}$ (square root). You need to understand what each of these symbols means and how they are used so we shall take a brief look at those with which you might not already be familiar.

4.3.1 Indices

The number 4 is the same as 2×2, that is, 2 multiplied by itself. We can write (2×2) as 2^2. In words, we would call this 'two raised to the power two' or simply 'two squared'. Thus:

$$2 \times 2 = \mathbf{2^2}$$

By similar reasoning we can say that:

$$2 \times 2 \times 2 = \mathbf{2^3}$$

and

$$2 \times 2 \times 2 \times 2 = \mathbf{2^4}$$

and so on.

In these examples, the number that we have used (i.e. 2) is known as the *base* whilst the number that we have raised it to is known as an *index*. Thus, 2^4 is called 'two to the power of four', and it consists of a base of 2 and an index of 4. Similarly, 5^3 is called 'five to the power of three' and has a base of 5 and an index of 3. Special names are used when the indices are 2 and 3, these being called 'squared' and 'cubed', respectively. Thus 7^2 is called 'seven squared' and 9^3 is called 'nine cubed'. When no index is shown, the power is 1, that is, 2^1 means 2. Also, note that *any* number raised to the power 0, is 1. Hence, $2^0 = 1$, $3^0 = 1$, $4^0 = 1$ and so on.

Example 4.6

Find the value of $2^5 + 3^3$
Now $2^5 = 2 \times 2 \times 2 \times 2 \times 2 = 32$ and $3^3 = 3 \times 3 \times 3 = 27$
So $2^5 + 3^3 = 32 + 27 = \mathbf{59}$

Example 4.7

Find the value of $10^2 - 5^3$
Now $10^2 = 10 \times 10 = 100$ and $5^3 = 5 \times 5 \times 5 = 125$
So $10^2 - 5^3 = 100 - 125 = \mathbf{-25}$

4.3.2 Reciprocals

The *reciprocal* of a number is when the index is -1 and its value is given by 1 divided by the base. Thus the reciprocal of 2 is 2^{-1} and its value is ½ or 0.5. Similarly, the reciprocal of 4 is 4^{-1} which means ¼ or 0.25.

Example 4.8

Find the value of $3^2 + 2^{-1}$
Now $3^2 = 3 \times 3 = 9$ and $2^{-1} = $ ½ or 0.5
So $3^2 + 2^{-1} = 9 + 0.5 = \mathbf{9.5}$

Example 4.9

Find the value of $\dfrac{1}{2} + \dfrac{1}{4} + \dfrac{1}{8}$

Now $\dfrac{1}{2} = 0.5$, $\dfrac{1}{4} = 0.25$, and $\dfrac{1}{8} = 0.125$

So $\dfrac{1}{2} + \dfrac{1}{4} + \dfrac{1}{8} = 0.5 + 0.25 + 0.125 = \mathbf{0.875}$

4.3.3 Negative indices

We have already said that the reciprocal of a number is the same as that number raised to the power -1. If the reciprocal happens to be a number raised to a power other than 1 then this is the same as the number raised to the same but *negative* power. This is probably sounding a lot more complex than it really is so here are a few examples:

$$\frac{1}{2} = 2^{-1}, \quad \frac{1}{2^2} = 2^{-2} \text{ and } \frac{1}{2^3} = 2^{-3}$$

Example 4.10

Find the value of 2^{-3}

Now $2^{-3} = \dfrac{1}{2^3} = \dfrac{1}{2 \times 2 \times 2} = \dfrac{1}{8} = \mathbf{0.125}$

4.3.4 Square roots

The *square root* of a number is when the index is ½. The square root of 2 is written as $2^{1/2}$ or $\sqrt{2}$. The value of a square root is the value of the base which when multiplied by itself gives the number. Since $3 \times 3 = 9$, then $\sqrt{9} = 3$. However, $(-3) \times (-3) = 9$, so we have a second possibility, that is, $\sqrt{9} = \pm 3$. There are always two answers when finding the square root of a number and we can indicate this is by placing a $\pm$ sign in front of the result meaning 'plus or minus'. Thus:

$4^{1/2} = \sqrt{4} = \pm 2$

and

$9^{1/2} = \sqrt{9} = \pm 3.$

Example 4.11

Find the value of $\sqrt{25}$
Now $25 = 5 \times 5$ (or -5×-5)
So $\sqrt{25} = \mathbf{\pm 5}$

Example 4.12

Find the value of $\sqrt{100} + \sqrt{4}$
Now $\sqrt{100} = \pm 10$ and $\sqrt{4} = \pm 2$
So $\sqrt{100} + \sqrt{4} = \pm 10 \pm 2$

Thus we have four potential answers:

$+10 + 2 = \mathbf{12}$
$+10 - 2 = \mathbf{8}$
$-10 + 2 = \mathbf{-8}$
$-10 - 2 = \mathbf{-12}$

Example 4.13

Find the value of $\dfrac{64^{1/2}}{16^{1/2}}$

Now $64^{1/2} = \sqrt{64} = 8$ and $64^{1/2} = \sqrt{16} = 4$

So $\dfrac{64^{1/2}}{16^{1/2}} = \dfrac{8}{4} = \mathbf{2}$

4.4 Electronic calculators

You will find that an electronic calculator (see Figure 4.3) can be extremely useful when it comes to solving engineering problems. You will need to ensure that the calculator has a full range of mathematical functions $(+, -, \div, \times, \sqrt{\ }, \text{etc.})$ as well the ability to use engineering

Figure 4.3 *A typical scientific calculator used for solving engineering problems*

notation (see later). Such calculators are often referred to as *scientific calculators*. The problems that follow all require the use of a calculator.

4.4.1 Significant figures

If you have tried Test your knowledge 4.6 you will probably have noticed that your answers were not *exactly* the same as mine! (Refer to Answers to numerical test your knowledge questions.) The reason for this is that I may have been working with more or less figures in my answer. In fact, I have not included more than five digits in any of my answers. Instead, I have *rounded up* or *rounded down* to the nearest digit in the least significant (rightmost) position. The rule is that, starting with the rightmost digit, if the number in a particular digit position is 5, or greater, then 1 is added to the number in the next (more significant) position. However, if the number in a particular digit position is less than 5, then the number in the next position remains unchanged. The number is then truncated at the point at which it has the correct number of digits (figures). This sounds a lot more complicated than it really is so here are a few examples given in Table 4.6.

Number	To four significant figures	To three significant figures	To two significant figures	To one significant figure
2.3333	2.333	2.33	2.3	2
3.6666	3.667	3.67	3.7	4
1.5923	1.592	1.59	1.6	2
6.8744	6.874	6.87	6.9	7
1638.4	1638	1640	1600	2000

Table 4.6 *Using the rounding up and rounding down method*

Notice in this last example how we have replaced some of the original digits with zeros!

Example 4.14

Find, to six significant figures, the square root of 251.
Using a calculator I find that:

$$\sqrt{251} = 15.84297952$$

The first six figures of this result are 15.8429. The next (seventh) digit is 7. This is larger than 5 so we must round up the last digit of our result by adding one to it. This makes our result (to six significant figures), **15.8430**.

4.4.2 Decimal places

Sometimes we are more interested in the number of digits after (i.e. to the right) of the decimal point rather than the total number of digits (figures) in an answer. In this case we simply round up or round down (as before) but ensure that we have the correct number of digits to the right of the decimal point. Any number to the left of the decimal point remains unchanged.

Example 4.15

Find, to three decimal places, the square root of 200.
Using a calculator I find that:

$$\sqrt{200} = 14.142315$$

The digits to the right of the decimal point must be reduced to just three. The fourth digit after the decimal point is 3; thus we do not need to round up the number to the left. This makes our result (to three decimal places) **14.142**.

4.5 Approximation

As precise calculations can take time (and may not always be required) engineers frequently need to make quick estimates based on *approximate values*. It is also quite useful to make a rough estimate of an answer to a problem or calculation before you arrive at a more accurate answer using a calculator. If there is a big discrepancy between the approximate result and the calculated answer you will know that something must have gone wrong and you will need to check your working again!

You may sometimes hear approximate values referred to as 'ball-park figures'. Approximation is frequently used when we do not actually need an exact figure straight away (we can always work this out later). Instead, we just need to get a 'feel' for what the exact value would be.

Let us suppose that we need to estimate the cost of a rectangular sheet of metal with sides 2.9 m and 4.1 m if we know that the metal costs £1.51 per m^2. To arrive at an approximate estimate of the cost we could simply round the numbers up or down and then multiply them together in order to find the area (in m^2) and then multiply the result by 1.5 (instead of 1.51) in order to arrive at an approximate cost in pounds.

Hence, estimated area of metal $= 3 \times 4 = 12\,m^2$ and
estimated cost $= 12 \times £1.5 = £18$

We can do all of this by applying a little mental arithmetic – there is actually no need to use a calculator!

Now, if we needed to arrive at an *exact* value, we could use the same reasoning but enter the values into a calculator (or use long multiplication) as follows:

Actual area of metal $= 2.9 \times 4.1 = 11.89\,m^2$ and
estimated cost $= 11.89 \times £1.51 = £17.954$

As you can see, our estimate of the cost was actually quite close to the real value but we got there much more quickly!

4.6 Variables and constants

Unfortunately, we do not always know the value of a particular quantity that we need to use in a calculation. In some cases the value might actually change, in which case we refer to it as a *variable*. In other cases, the value might be fixed (and we might actually prefer not to actually quote its value). In this case we refer to the value as a *constant*.

In either case, we can use a *symbol* to represent the quantity. The symbol itself (often a single letter) is a form of shorthand notation. For example, in the case of the voltage produced by a battery we would probably use v to represent *voltage* whereas, in the case of the time taken to travel a certain distance, we might use t.

An example of a *variable* quantity is the outside temperature. On a hot summer's afternoon the temperature may well exceed 30°C whilst, on a cold winter's morning it might be as little as −4°C. An example of a *constant* quantity might be the temperature at which water freezes and becomes ice; that is, 0°C.

Let us take another example. Figure 4.4 shows a lorry carrying several different loads. In Figure 4.4(a) the lorry is not carrying any load at all. The weight of the lorry in this condition (known as its *unladen weight*) is 4 tonnes. In Figure 4.4(b) a load of 1 tonne is being carried. If you add this to the unladen weight of the lorry the total weight of the lorry is $4 + 1$ or 5 tonnes. In Figure 4.4(c)–(e) loads of 2.5, 4 and 8 tonnes are being carried and this results in total weights of 6.5, 8 and 10 tonnes, respectively.

It should be obvious that the unladen weight is a *constant* and the load weight is a *variable*. We can express the relationship between the unladen weight, load weight and total weight using a simple formula, as follows:

Total weight = unladen weight + load weight

$$W_t = W_u + W_l$$

where W_t represents the total weight, W_u represents the unladen weight, and W_l represents the load weight.

The formula can be quite useful. For example, suppose the lorry manufacturer has specified a maximum total weight of 12.5 tonnes. We might want to check that we do not exceed this value. We can calculate the maximum load weight by re-arranging the formula as follows:

$$W_l = W_t - W_u$$

where $W_t = 12.5$ tonnes and $W_u = 4$ tonnes.

From which:

$$W_l = 12.5 - 4 = 8.5 \text{ tonnes}$$

Hence the maximum load that the lorry can carry (without exceeding the manufacturer's specification) is 8.5 tonnes.

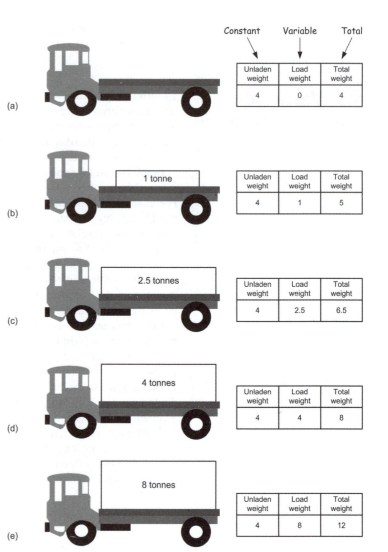

Figure 4.4 *A lorry with various different loads (all weights in tonnes)*

4.7 Proportionality

In engineering applications, when one quantity changes it normally affects a number of other quantities. For example, if the engine speed of a car increases its road speed will also increase (see Figure 4.5). To put this in a mathematical way we can say that:

road speed is *directly proportional* to engine speed

Using mathematical notation and symbols to represent the quantities, we would write this as follows:

$$v \propto N$$

where v represents road speed and N represents the engine speed.

In some cases, an increase in one quantity produces a *reduction* in another quantity.

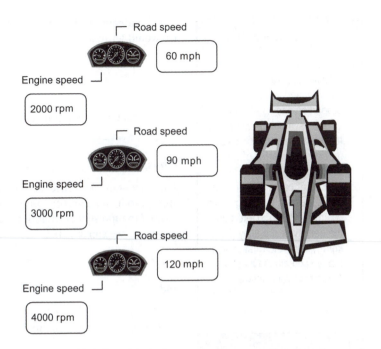

Figure 4.5 *Relationship between engine speed and road speed*

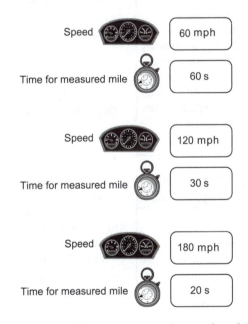

Figure 4.6 *Relationship between road speed and time taken to travel a measured distance*

For example, if the road speed of a car increases the time taken for it to travel a measured distance will decrease (see Figure 4.6). To put this in a mathematical way we would say that:

time taken to travel a measured distance is *inversely proportional* road speed

1. The density of a body is directly proportional to its mass and inversely proportional to its volume. Use the symbols ρ, M and V, to write down an expression for density in terms of mass and volume.
2. A block of polyurethane foam has a mass of 0.155 kg and a volume of 1.2 m³. Determine the density of the alloy (in kg/m³) using the relationship that you obtain in Question 1. Express your answer correct to four decimal places.

Key point

If the value of one variable increases when the value of another variable increases we say that they are *proportional* to one another. If the value of one variable increases when the value of another variable decreases we say that they are *inversely proportional* to one another.

Test your knowledge 4.12

Which of the following expressions are true?

(a) $\dfrac{2^2}{2} = 2$

(b) $\dfrac{1}{2^{-1}} = 2$

(c) $4^{1/2} = 2$

(d) $\dfrac{(2^2)^2}{2^3} = 2$

(e) $\dfrac{2^{10} \times 2^2}{2^{11}} = 2$

Using mathematical notation and symbols to represent the quantities, we would write this as follows:

$$t \propto \frac{1}{v}$$

where t represents the time taken and v represents the road speed.

Example 4.16

The current in an electric circuit is directly proportional to the voltage applied to it and inversely proportional to the resistance of the circuit. Using I to represent current, V to represent voltage, and R to represent resistance we can say that:

$I \propto V$ (current, I, is proportional to voltage, V)

and

$I \propto \dfrac{1}{R}$ (current, I, is inversely proportional to resistance, V)

We can combine these two relationships to obtain an *equation* which involves all three variables, I, V and R:

$$I = \frac{V}{R}$$

We shall explore this important relationship (which comes from Ohm's law) later on in this chapter.

Example 4.17

The power delivered to a loudspeaker is proportional to the square of the voltage applied to it and inversely proportional to the impedance of the speaker. Determine the power that would be delivered to a 4 ohm (Ω) loudspeaker when connected to an amplifier that delivers 20 V.

From the information given, and using P, V and Z to represent power, voltage and impedance, we can obtain the following relationships:

$P \propto V^2$ (power, P, is proportional to the square of the voltage, V)

and

$P \propto \dfrac{1}{Z}$ (power, P, is inversely proportional to the impedance, Z)

We can combine these two relationships to obtain an *equation*:

$$P = \frac{V^2}{Z}$$

Now we know that $V = 20\,V$ and $Z = 4\,\Omega$. Replacing the symbols that we have been using by the values that we know gives:

$$P = \frac{V^2}{Z} = \frac{20^2}{4} = \frac{400}{4} = \mathbf{100\,W}$$

4.8 Laws of indices

When simplifying calculations involving indices, certain basic rules or laws called the *laws of indices* can be applied. These are listed below:

- when multiplying two or more numbers having the same base, the indices are added. Thus $2^2 \times 2^4 = 2^{2+4} = 2^6$.

- when a number is divided by a number having the same base, the indices are subtracted. Thus $2^5/2^2 = 2^{5-2} = 2^3$.

- when a number which is raised to a power is raised to a further power, the indices are multiplied. Thus $(2^5)^2 = 2^{5\times2} = 2^{10}$. when a number has an index of 0, its value is 1. Thus $2^0 = 1$.

- when a number is raised to a negative power, the number is the reciprocal of that number raised to a positive power. Thus $2^{-4} = 1/2^4$. Similarly, $1/2^{-3} = 2^3$.

- when a number is raised to a fractional power the denominator of the fraction is the root of the number and the numerator is the power. Thus $4^{3/4} = \sqrt[4]{4^3} = (2)^{3/2} = 2\sqrt{2}$ and $25^{1/2} = \sqrt{25^1} = \pm5$.

Key point

When multiplying two numbers having the same base the indices are added. If one number is divided by another of the same base, the indices are subtracted. Thus $3^3 \times 3^2 = 3^5$ and $3^3/3^2 = 3^1 = 3$.

Key point

When a number has an index of 0 its value is 1 (regardless of the actual value of the number). Thus 2^0, 3^0, 4^0 and so on are all equal to 1.

Example 4.18

Find the value of $\dfrac{3^3 \times 3^4}{3^5}$

Now $\dfrac{3^3 \times 3^4}{3^5} = \dfrac{3^{3+4}}{3^5} = \dfrac{3^7}{3^5} = 3^{7-5} = 3^2 = \mathbf{9}$

4.9 Standard form

A number written with one digit to the left of the decimal point and multiplied by 10 raised to some power is said to be written in *standard form* (this is also sometimes referred to as *scientific notation*) Thus: 1234 is written as 1.234×10^3 in standard form, and 0.0456 is written as 4.56×10^{-2} in standard form.

When a number is written in standard form, the first factor is called the *mantissa* and the second factor is called the *exponent*. Thus the number 6.8×10^3 has a mantissa of 6.8 and an exponent of 10^3.

Numbers having the same exponent can be added or subtracted in standard form by adding or subtracting the mantissae and keeping the exponent the same. Thus:

$$2.3 \times 10^4 + 3.7 \times 10^4 = (2.3 + 3.7) \times 10^4 = 6.0 \times 10^4$$

and

$$5.7 \times 10^{-2} - 4.6 \times 10^{-2} = (5.7 - 4.6) \times 10^{-2} = 1.1 \times 10^{-2}$$

When adding or subtracting numbers it is quite acceptable to express one of the numbers in non-standard form, so that both numbers have the same exponent. This makes things much easier as the following example shows:

$$2.3 \times 10^4 + 3.7 \times 10^3 = 2.3 \times 10^4 + 0.37 \times 10^4$$
$$= (2.3 + 0.37) \times 10^4$$
$$= 2.67 \times 10^4$$

Alternatively,

$$2.3 \times 10^4 + 3.7 \times 10^3 = 23,000 + 3700 = 26,700 = 2.67 \times 10^4$$

The laws of indices are used when multiplying or dividing numbers given in standard form. For example,

$$(2.5 \times 10^3) \times (5 \times 10^2) = (2.5 \times 5) \times (10^{3+2})$$
$$= 12.5 \times 10^5 \text{ or } 1.25 \times 10^6$$

To round this section off, here are some examples of engineering quantities expressed in standard form:

$$221\,N = 2.21 \times 10^2\,N$$
$$0.035\,V = 3.5 \times 10^{-2}\,V$$
$$454\,kHz = 4.54 \times 10^5\,Hz$$
$$65.5\,\mu C = 6.55 \times 10^{-5}\,C$$

4.9.1 Engineering notation

Engineering notation is very similar to standard form (scientific notation) but uses a mantissa that lies in the range 1–999 and an index that is a multiple of three. Engineering notation is useful because it is very easy to convert to the standard multiples and sub-multiples of the standard units (which are all power of three). Here are some examples of using engineering notation:

$$3495\,m = 3.495 \times 10^3\,m\ (=3.495\,km)$$
$$0.075\,V = 75 \times 10^{-3}\,V\ (=75\,mV)$$
$$12,576\,N = 12.576 \times 10^3\,N\ (=12.576\,kN)$$
$$67,625,000\,Hz = 67.625 \times 10^6\,Hz\ (=67.625\,MHz)$$
$$0.00025\,A = 250 \times 10^{-6}\,A\ (=250\,\mu A)$$

4.9.2 Exponent notation

Unfortunately, computers and electronic calculators are not well suited to the entry of numbers expressed using either standard form or using engineering notation. To overcome this problem, scientific and engineering calculators make use of a special mode of data entry using a key which is often labelled 'EXP' (see Figure 4.3) and a display which uses an 'E' symbol to indicate that *exponent notation* is being used.

In exponent notation, the 'E' means 'multiply by 10 raised to a power given by the number that follows. So, '1E1' means $1 \times 10^1 = 10$,

Test your knowledge 4.16

1. Express (a) 19,950 and (b) 0.0075 and using in exponent notation.
2. Express (a) 1.59125E2 and (b) 1.915E−6 using standard form.

'1E2' means $1 \times 10^2 = 100$, '1E3' means $1 \times 10^3 = 1000$ and so on. Hence, instead of writing (or entering) 10,000 we can simply write '1E4' (in other words 1×10^4). Similarly, we would write 1×10^5 for 100,000 and so on.

Here are some examples:

$$2251 = 2.251E3$$

$$68,295 = 6.8295E4$$

$$1,577,625 = 1.577625E6$$

$$0.005491 = 5.491E{-3}$$

$$0.00001221 = 1.221E{-5}$$

4.10 Equations

When we need to understand the relationship between different quantities we often express this in the form of an *equation*. To save time and effort we write an equation using symbols rather than words, for example, instead of writing 'distance travelled is the same as speed divided by time' we would simply write:

$$s = v \times t$$

where s represents distance (in metres), v represents speed (or velocity) (in metres per second), and t represents time (in seconds).

So, if we need to find the distance travelled by a car travelling at 45 m/s in 20 s we could substitute values in the equation, using $v = 10$ and $t = 20$, as follows:

$$s = v \times t = 45 \times 20 = \textbf{900 m}$$

In fact, engineers frequently need to solve equations in order to find the value of an unknown quantity, as the following examples show.

Example 4.19

A current of 0.5 A flows in a 56 Ω resistor. Given that $V = IR$ determine the voltage that appears across the resistor.

It is a good idea to get into the habit of writing down what you know before attempting to solve an equation. In this case:

$I = 0.5$ A
$R = 56\,\Omega$
$V = ?$

Now

$$V = IR = 0.5 \times 56 = \textbf{28 V}$$

Example 4.20

Resistors of 3.9, 5.6 and 10 kΩ are connected in parallel. Calculate the effective resistance of the circuit.

The resistance of a parallel circuit is given by the equation:

$$\frac{1}{R} = \frac{1}{R1} + \frac{1}{R2} + \frac{1}{R3}$$

Now we know that:

$$R1 = 3.9\,\text{k}\Omega = 3.9 \times 10^3\,\Omega$$
$$R2 = 5.6\,\text{k}\Omega = 5.6 \times 10^3\,\Omega$$
$$R3 = 10\,\text{k}\Omega = 1 \times 10^4\,\Omega = 10 \times 10^3\,\Omega$$

Hence:

$$\frac{1}{R} = \frac{1}{3.9 \times 10^3} + \frac{1}{5.6 \times 10^3} + \frac{1}{10 \times 10^3}$$

$$= \frac{10^{-3}}{3.9} + \frac{10^{-3}}{5.6} + \frac{10^{-3}}{10} = \left(\frac{1}{3.9} + \frac{1}{5.6} + \frac{1}{10}\right) \times 10^{-3}$$

$$= (0.256 + 0.179 + 0.1) \times 10^{-3} = 0.535 \times 10^{-3}$$

Now since $1/R = 0.535 \times 10^{-3}$ we can invert both sides of the equation so that:

$$R = \frac{1}{0.535 \times 10^{-3}} = 1.87 \times 10^3 = \mathbf{1.87\,k\Omega}$$

Any arithmetic operation may be applied to an equation as long as the equality of the equation is maintained. In other words, the *same* operation *must* be applied to both the left-hand side (LHS) and the right-hand side (RHS) of the equation.

Example 4.21

A copper wire has a length l of 1.5 km, a resistance R of 5 Ω and a resistivity ρ of $17.2 \times 10^{-6}\,\Omega$ mm. Find the cross-sectional area, a, of the wire, given that $R = \rho l/a$.

Once again, it is worth getting into the habit of summarising what you know from the question and what you need to find (do not forget to include the units):

$$R = 5\,\Omega$$
$$\rho = 17.2 \times 10^{-6}\,\Omega\,\text{mm}$$
$$l = 1500 \times 10^3\,\text{mm}$$
$$a = ?$$

Since $R = \rho l/a$ then

$$5 = \frac{(17.2 \times 10^{-6})\,(1500 \times 10^3)}{a}$$

Cross multiplying (i.e. exchanging the '5' for the 'a') gives:

$$a = \frac{(17.2 \times 10^{-6})\,(1500 \times 10^3)}{5}$$

Now group the numbers and the powers of 10 as shown below:

$$a = \frac{17.2 \times 10^{-6} \times 1500 \times 10^3}{5}$$

Next simplify as far as possible:

$$a = \frac{17.2 \times 1500}{5} \times 10^{-6+3}$$

Finally, evaluate the result using your calculator:

$$a = 5160 \times 10^{-3} = 5.16$$

Since we have been working in mm, the result, a, will be in mm^2. Hence:

$$a = \textbf{5.16 mm}^2$$

It is worth noting from the previous example that we have used the laws of indices to simplify the powers of 10 before attempting to use the calculator to determine the final result. The alternative to doing this is to make use of the exponent facility on your calculator. Whichever technique you use it is important to be confident that you are correctly using the exponent notation since it is not unknown for students to produce answers that are incorrect by a factor of 1,000 or even 1,000,000 and an undetected error of this magnitude could be totally disastrous!

Before attempting to substitute values into an equation, it is important to be clear about what you know (and what you do not know) and always make sure that you have the correct units.

Example 4.22

The reactance, X, of a capacitor is given by $X = 1/2\pi f C$.

Find the value of capacitance that will exhibit a reactance of $10\,\text{k}\Omega$ at a frequency of $400\,\text{Hz}$.

First, to summarise what we know:

$X = 10\,\text{k}\Omega = 10 \times 10^3\,\Omega$
$f = 400\,\text{Hz}$
$\pi = 3.142$ (or use the 'π' button on your calculator!)
$C = ?$

We need to re-arrange the formula $X = 1/2\pi f C$ to make C the subject. This is done as follows:

Cross multiplying gives:

$$C = \frac{1}{2\pi f \times X}$$

(notice how we have effectively 'swapped' the C and the X over)

Next, replacing π, C and f by the values that we know gives:

$$C = \frac{1}{2 \times 3.142 \times 400 \times 10 \times 10^3} = \frac{1}{25{,}136 \times 10^3}$$

$$= \frac{1}{2.5136 \times 10^7} = \frac{1}{2.5136} \times 10^{-7} = 0.398 \times 10^{-7}\,\text{F}$$

Finally, it would be sensible to express the answer in nF (rather than F). To do this, we simply need to multiply the result by 10^9, as follows:

$$C = 0.398 \times 10^{-7} \times 10^9 = 0.398 \times 10^{9-7}$$
$$= 0.398 \times 10^2 = \mathbf{39.8\,nF}$$

Example 4.23

The frequency of resonance, f, of a tuned circuit is given by the relationship:

$$f = \frac{1}{2\pi\sqrt{LC}}$$

where f is the frequency (in Hz), L is the inductance (in H) and C is the capacitance (in F). If a tuned circuit is to be resonant at 6.25 MHz and $C = 100\,pF$, determine the value of inductance, L.

Here we know that:

$$C = 100\,pF = 100 \times 10^{-12}\,F$$
$$f = 6.25\,MHz = 6.25 \times 10^6\,Hz$$
$$\pi = 3.142 \text{ (or use the '}\pi\text{' button on your calculator!)}$$
$$L = ?$$

First we will re-arrange the formula $f = 1/2\pi\sqrt{LC}$ in order to make L the subject:

Squaring both sides gives:

$$f^2 = \left(\frac{1}{2\pi\sqrt{LC}}\right)^2 = \frac{1^2}{2^2\pi^2\left(\sqrt{LC}\right)^2} = \frac{1}{4\pi^2 LC}$$

Re-arranging gives:

$$L = \frac{1}{4\pi^2 f^2 C}$$

We can now replace f, C and π by the values that we know:

$$L = \frac{1}{4 \times 3.142^2 \times (6.25 \times 10^6)^2 \times 100 \times 10^{-12}}$$
$$= \frac{1}{39.49 \times 39.06 \times 10^{12} \times 100 \times 10^{-12}}$$
$$= \frac{1}{154{,}248} = \frac{1}{1.542{,}48 \times 10^5} = \frac{1}{1.542{,}48} \times 10^{-5}$$
$$= 0.648 \times 10^{-5} = 6.48 \times 10^{-6} = \mathbf{6.48\,\mu H}$$

4.11 Length and area

Length and area are extremely important in engineering and engineers frequently have to carry out calculations involving these quantities. Measurement of length is often carried out with a rule but when we need more accurate measurements, particularly of small components, we often use a vernier caliper (see Figure 4.7).

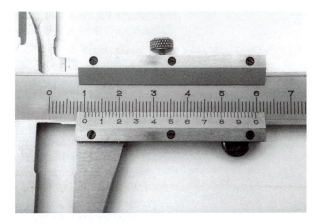

Figure 4.7 *A vernier caliper used for accurate measurement of small components. The reading indicated is 1.04*

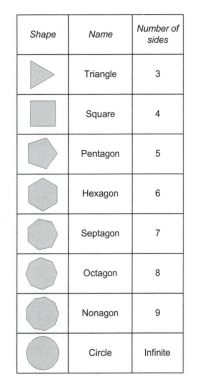

Shape	Name	Number of sides
	Triangle	3
	Square	4
	Pentagon	5
	Hexagon	6
	Septagon	7
	Octagon	8
	Nonagon	9
	Circle	Infinite

Figure 4.8 *Some common shapes*

Area and volume can be calculated from measurements of length but, as we shall see later in this section, the way that we calculate area or volume depends on the shape that we are dealing with.

4.11.1 Shapes

Figure 4.8 shows several common shapes which have different numbers of straight sides. Notice that the triangle has three sides, the square has four sides, the pentagon has five sides and so on. And, although the circle does not have straight lines, it can actually be considered to be an object with an infinite number of sides of equal length!

4.11.2 Angular and linear measure

Being able to measure angles as well as length is important in many engineering applications. The essential difference between *angular measure* and *linear measure* is illustrated in Figure 4.9. In Figure 4.9(a) we are concerned with the distance between points A and B measured along a straight line which joins the two points. In Figure 4.9(b), we are concerned with the amount of rotation between lines A and B (which can be of any length) which meet at point O.

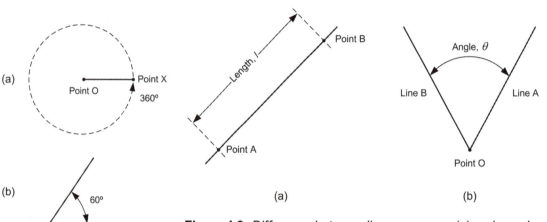

Figure 4.9 *Difference between linear measure (a) and angular measure (b)*

One complete rotation, starting at point X and returning to point X in Figure 4.10(a), is equivalent to an angle of 360°. Some other angles are illustrated in Figure 4.10(b)–(d).

When lines are perpendicular to one another (i.e. at right angles) the angle between them is 90°, as shown in Figure 4.11.

4.11.3 Area

The formulae for determining areas of various shapes are shown in Figure 4.12.

Some simple rectangular shapes are shown in Figure 4.13. In Figure 4.13(a) the shape is a square measuring 1 m × 1 m. The area of the shape is thus $1 \times 1 = 1\,\text{m}^2$. In Figure 4.13(b) the shape is a square measuring 2 m × 2 m. The area of the shape is thus $2 \times 2 = 4\,\text{m}^2$. In Figure 4.13(c) the shape is a square measuring 3 m × 3 m. The area of the shape is thus $3 \times 3 = 9\,\text{m}^2$.

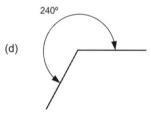

Figure 4.10 *(a) One complete rotation is equivalent to moving through an angle of 360°; (b)–(d) some other angles*

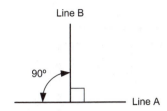

Figure 4.11 *Perpendicular lines*

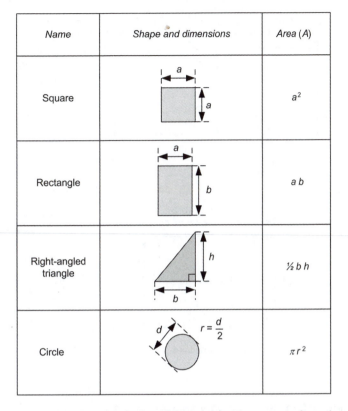

Figure 4.12 *Formulae for determining the area of various engineering shapes*

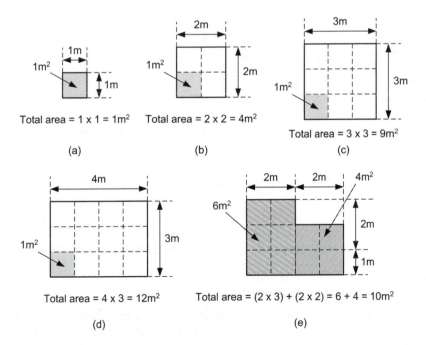

Figure 4.13 *Area of rectangular shapes*

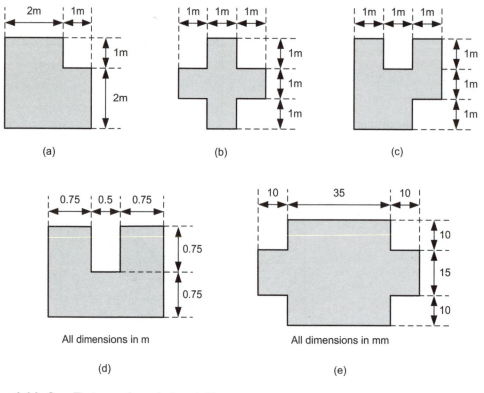

(a)   (b)   (c)

All dimensions in m   All dimensions in mm

(d)   (e)

Figure 4.14 *See Test your knowledge 4.18*

Test your knowledge 4.18

Find the area of each shape shown in Figure 4.14.

In Figure 4.13(d) we are dealing with a rectangle (rather than a perfect square). The rectangle has dimensions 4 m × 3 m and its area is $4 \times 3 = 12\,\text{m}^2$. Finally, in Figure 4.13(e) we have a shape that can be divided into two rectangles. The dimensions of one rectangle is 2 m × 3 m whilst the other is a perfect square measuring 2 m × 2 m. The area of the shape is the sum of the areas of the two rectangles, that is, $(2 \times 3) + (2 \times 2) = 6 + 4 = 10\,\text{m}^2$.

4.11.4 Triangles

As you saw in Figure 4.8, a triangle is an object that has three sides. The sum of the interior angles of a triangle is 180°. Thus, if we know two of the angles of a triangle we can find the remaining (third) angle.

Example 4.24

Determine the third angle of the triangle shown in Figure 4.15.
    Now the sum of the angles in a triangle is 180°. Hence:

$\theta + 45° + 60° = 180°$

Figure 4.15 *See Example 4.24*

So,

$$\theta = 180° - (45° + 60°) = 180° - 105° = \mathbf{75°}$$

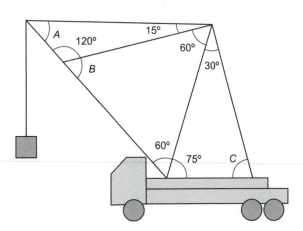

Figure 4.16 *See Test your knowledge 4.19*

Some shapes involving triangles are shown in Figure 4.17. In Figure 4.17(a) the shape is a triangle which is half of a perfect square having sides $1\,\text{m} \times 1\,\text{m}$. The area of the triangle is thus $0.5 \times (1 \times 1) = 0.5\,\text{m}^2$. In Figure 4.17(b) the area is made up from three perfect squares (each of area $1\,\text{m}^2$) and one triangle having an area of $0.5\,\text{m}^2$. The total area is thus $3.5\,\text{m}^2$.

Finally, the shape in Figure 4.17(c) can be divided into a rectangle with an area of $12\,\text{m}^2$ and a triangle having an area of $3\,\text{m}^2$. The total area is thus $(12 + 3) = 15\,\text{m}^2$.

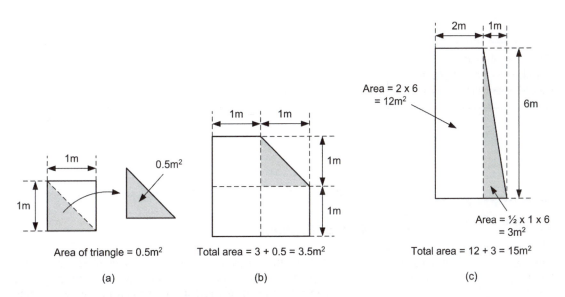

Figure 4.17 *Some shapes involving triangles*

Test your knowledge 4.20

Find the area of each shape shown in Figure 4.18.

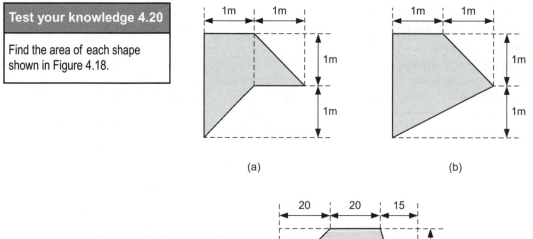

(a)   (b)

(c)

All dimensions in mm

Figure 4.18 *See Test your knowledge 4.20*

Test your knowledge 4.21

Determine the total area of the tail plane for the *Gulfstream* aircraft shown in Figure 4.19.

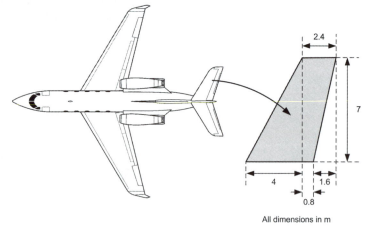

All dimensions in m

Figure 4.19 *See Test your knowledge 4.21*

Key point

The sum of the angles of a triangle is 180°. This, if two of the angles are known the third can easily be found.

4.11.5 Circles

The circle is another shape that is often found in engineering and you should be able to think of quite a few applications of this particular shape! The properties of a circle are illustrated in Figure 4.20 and summarised in Table 4.7.

Property	Relationship
Radius, r	$r = d/2$
Diameter, d	$d = 2r$
Circumference, l	$l = \pi d$
Area, A	$A = \pi r^2$

Table 4.7 *The property of a circle*

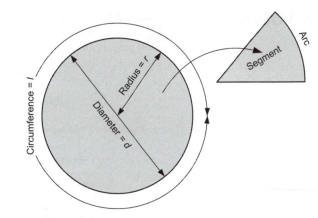

Figure 4.20 *Properties of a circle*

Example 4.25

A circular test track is to be constructed. If the test track is to have a circumference of 2.5 km, determine the required diameter of the circle.

The relationship between the circumference, l, of a circle and its diameter, d, is as follows:

$$l = \pi d$$

Re-arranging this in order to make d the subject of the equation gives:

$$d = \frac{l}{\pi}$$

and since $l = 2.5 \times 10^3$ m we have:

$$d = \frac{2.5 \times 10^3}{\pi} = \frac{2.5 \times 10^3}{3.142} = 0.796 \times 10^3 = \textbf{796 m}$$

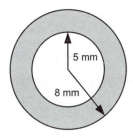

Figure 4.21 *See Example 4.26*

Example 4.26

Find the surface area of the washer shown in Figure 4.21.

The area of the washer can be found by subtracting the area of a circle having a radius of 5 mm from a circle that has a radius of 8 mm.

Area of a circle having a radius of 8 mm:

$$A_1 = \pi r^2 = \pi \times (8 \times 10^{-3})^2 = \pi \times 64 \times 10^{-6} = 201.06 \times 10^{-6} \text{m}^2$$

Area of a circle having a radius of 5 mm:

$$A_2 = \pi r^2 = \pi \times (5 \times 10^{-3})^2 = \pi \times 25 \times 10^{-6} = 78.54 \times 10^{-6} \text{m}^2$$

Thus the area of the washer is given by:

$$A = A_1 - A_2 = (201.06 - 78.54) \times 10^{-6} = \textbf{122.52} \times \textbf{10}^{-6}\,\textbf{m}^2$$

4.12 Volume

Area is a *two-dimensional* (2D) property which is found my multiplying one length by another length whilst volume is a *three-dimensional* (3D) property found my multiplying a *cross-sectional area* by a length. Figure 4.22 illustrates this important concept.

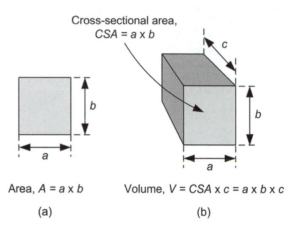

Figure 4.22 *Area and volume*

The volumes of some common 3D objects are listed in Table 4.8.

Object	Formula
Cube with side length $= s$	$V = s^3$
Rectangular block with dimensions a, b, c	$V = abc$
Cylinder with height h and radius r	$V = \pi r^2 h$
Sphere with radius r	$A = 4/3\pi r^3$

Table 4.8 *Volumes of common 3D objects*

Example 4.27

Determine the volume of the trailer shown in Figure 4.23.

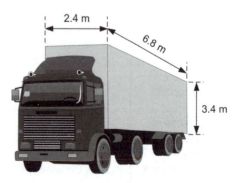

Figure 4.23 *See Example 4.27*

Now the volume of a rectangular box is given by:

$$V = a \times b \times c$$

and from Figure 4.23:

$$a = 2.4\,\text{m}, b = 6.8\,\text{m}, \text{and } c = 3.4\,\text{m}$$

Hence,

$$V = 2.4 \times 6.8 \times 3.4 = \mathbf{55.5\,m^2}$$

Example 4.28

Determine the volume (in m^3) of a cylindrical block of alloy having a diameter of 24 mm and a height of 120 mm.

Now the volume of a cylinder is given by:

$$V = \pi r^2 h$$

Since the diameter, $d = 24\,\text{mm}$, the radius, r, will be given by:

$$r = \frac{d}{2} = \frac{24}{2} = \mathbf{12\,mm}$$

The volume, V, can now be calculated as follows:

$$V = \pi r^2 h = 3.142 \times (12)^2 \times 120 = 54{,}293.8\,\text{mm}^3$$

To convert this to m^3 we need to multiply by 1×10^{-9} as follows:

$$V = (5.42938 \times 10^3) \times 10^{-9} = \mathbf{5.42938 \times 10^{-6}\,m^3}$$

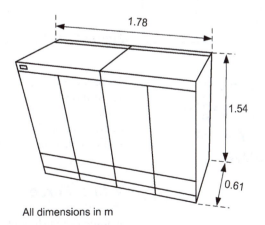

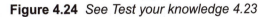
All dimensions in m

Figure 4.24 *See Test your knowledge 4.23*

Test your knowledge 4.23

Determine the internal volume of the equipment cabinet shown in Figure 4.24.

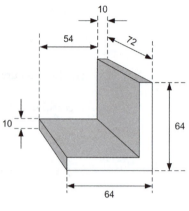

All dimensions in mm

Figure 4.25 *See Test your knowledge 4.24*

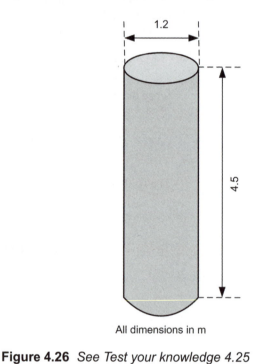

All dimensions in m

Figure 4.26 *See Test your knowledge 4.25*

4.13 Force, mass, weight, density and pressure

Force, mass, weight and density are important in many engineering applications and it is quite likely that you already have some idea of what these terms mean. However, do you understand the difference between mass and weight? And, how would measure a force? This section helps you get up to speed with these important concepts.

4.13.1 Force

A force is a push or pull exerted by one object on another and, if the object remains in *equilibrium* (i.e. if it does not move), for every force

Key point

When a body is at rest (or in *equilibrium*) the *action* and *reaction* forces acting on it will be equal and opposite.

Key point

If the action and reaction forces acting on a body are not equal and opposite, motion will be produced.

there is another equal and opposite force that acts against it. An applied force is often called an *action* whilst the opposing force is referred to as a *reaction*. As long as the object (or *body*) does not move, action and reaction are equal and opposite forces.

It also follows that, if the forces of action and reaction acting on an object are not equal and opposite, the object will move. You can test this theory out very easily by finding a wall and pushing against it. If the wall does not move (hopefully it will not!) then you will experience a force pushing back. If you increase the force that you apply to the wall (the *action*) the force pushing back (the *reaction*) will also increase by the same amount.

Now try the same experiment by pushing against a door that it partially open. There will still be some force exerted back by the door but this will be much less than the force that you apply. Due to this imbalance of forces (action being greater than reaction) the door will move and will swing open.

4.13.2 Mass

The mass of a body is defined as the quantity of matter in the body. The mass of a body remains the same regardless of where the body is. So, for example, a mass of 50 kg will be the same on the surface of the earth as it will be in outer space (where there is 'zero gravity').

4.13.3 Weight

The weight of a body is determined by its mass and the gravitational force acting on the body. So, if there is no gravitational force (e.g. in outer space) then a body will have no weight! However, in most practical cases we are concerned with what things weigh on the surface of the earth in which case the relationship between mass and weight is given by:

$$W = mg$$

where W is the weight in newton (N), and g is the *gravitational acceleration* (in m/s^2). On the surface of the earth g is a constant equal to 9.81 m/s^2.

The weight of a body decreases as the body is moved away from the centre of the Earth. Weight obeys the inverse square law. This simply means that weight is inversely proportional to the square of the distance from the centre of the earth. In other words:

$$W \propto \frac{1}{d^2}$$

where W is the weight (in N) and d is the distance from the centre of the Earth (in m).

Example 4.29

A light alloy beam has a mass of 14.5 kg. Determine the weight of the beam.

Here we will assume that the beam is being used at the Earth's surface. In which case:

$$W = mg = 14.5 \times 9.81 = \mathbf{142.25\,N}$$

Example 4.30

A remotely operated vehicle (ROV) weighing 4.25 kN on Earth, is to be used on a mission to explore the surface of Mars. Given that the gravitational acceleration on Mars is 0.38 of that on Earth, determine the weight of the ROV on Mars.

The weight of the ROV will be reduced in direct proportion to the reduction in gravitational acceleration. Hence the ROV will weigh $0.38 \times 4.25\,\text{kN} = \mathbf{1.615\,kN}$ on Mars.

4.13.4 Density

The density of a body is defined as the mass per unit volume. Expressing this as a formula gives:

$$\rho = \frac{m}{V}$$

where ρ is the density in kg/m^3, m is the mass in kg, and V is the volume in m^3.

We sometimes express the density of an object relative to that of pure water (at 4°C). The density of water under these conditions is 1000 kg/m^3. The density and relative density of various engineering materials are shown in Table 4.9.

Material	Density (kg/m³)	Relative density
Aluminium	2700	2.7
Brass	8500	8.5
Cast iron	7350	7.35
Concrete	2400	2.4
Copper	8960	8.96
Glass	2600	2.6
Mild steel	7850	7.85
Wood (oak)	690	0.69

Table 4.9 *Density and relative density of various engineering materials*

Example 4.31

Determine, to two decimal places, the mass of an aluminium block which has dimensions 50 mm × 110 mm × 275 mm.

The total volume of the aluminium block will be given by:

$$V = (50 \times 10^{-3}) \times (110 \times 10^{-3}) \times (275 \times 10^{-3})$$
$$= (1.5125 \times 10^{6}) \times 10^{-9} = \mathbf{1.5125 \times 10^{-3}\,m^3}$$

Key point

Mass and weight are not the same thing! The mass of an object remains the same wherever it is. The weight of a body is determined by the product of the body's mass and the gravitational force that is acting on it.

Key point

The density of an object is found by dividing its mass by its volume. The density of a particular material is a fundamental property of that material.

Test your knowledge 4.26

1. A vehicle chassis has a mass of 475 kg. Determine the weight of the chassis.
2. A uniform oak beam has a cross-sectional area of 0.06 m^2 and a length of 4.5 m. Determine the mass of the beam.
3. A mild steel girder has a mass of 560 kg and is supported by two concrete pillars each having a cross-sectional area of 0.044 m^2. Assuming that the load is distributed evenly, determine the pressure exerted on each of the two pillars.

Key point

The pressure acting on a surface is defined as the ratio of the force acting perpendicular to the surface to the area over which the force is acting.

Re-arranging the formula $\rho = m/V$ to make m the subject gives:

$$m = \rho \times V$$

From Table 4.9, the value of ρ for aluminium is 2700 kg/m^3 hence the mass of the block will be given by:

$$m = 2700 \times 1.5125 \times 10^{-3} = 4083.75 \times 10^{-3} = \textbf{4.08 kg}$$
(to two decimal places)

4.13.5 Pressure

Pressure is exerted whenever a force is applied to an object, such as a floor, wall or the inside surfaces of a container. Pressure is defined as the ratio of force (or load) applied perpendicular (i.e. at right angles) to the surface, to the area over which the force (or load) acts.

Thus:

$$P = \frac{F}{A}$$

where P is the pressure (in Pa), F is the force (in N) and A is the area (in m^2).

Example 4.32

A lathe weighs 425 kN. Determine the pressure exerted on the workshop floor if the load is distributed over a surface area of 0.875 m^2.

Now:

$$P = \frac{F}{A} = \frac{425 \times 10^3}{0.875} = \textbf{485.7 kPa}$$

4.14 Force and acceleration

Test your knowledge 4.27

1. An object having a mass of 125 kg is to have an acceleration of 15 m/s^2. What force is required to do this?
2. A force of 75 kN is applied to an object which has a mass of 17.5 kg. Determine the acceleration produced.

In the previous section we briefly introduced as a push or pull exerted by one object on another. We also explained how weight is a force that results from gravity. In fact, gravitational force is something that we all experience all of the time and it is what keeps us firmly in place on the surface of the Earth – without it we would simply drift off into space!

When forces of action and reaction acting on a body are not equal and opposite (i.e. when the body is not actually in a state of equilibrium) the body will begin to move. In fact, it will experience an *acceleration*.

The relationship between the mass of the body, the force applied, and the acceleration that is produced is very important. If we want something to move fast we need to know how much force to apply! It should also be fairly obvious that, the larger the mass of the body the more force we would need to apply in order to make it move.

In fact, the unit of force (the newton) is usually defined in terms of this important relationship. The definition is as follows:

A force of one Newton is that which would produce an acceleration of one metre per square second in a mass of one kilogram.

The relationship is as follows:

$$F = m \times a$$

where F is the force (in N), m is the mass (in kg), and a is the acceleration (in m/s^2).

Example 4.33

A rocket produces a thrust of 9.75 kN. If the rocket has a mass of 42.4 kg, determine the acceleration produced.

Re-arranging $F = m \times a$ to make a the subject gives:

$$a = \frac{F}{m} = \frac{9.75 \times 10^3}{42.4} = \mathbf{230\,m/s^2}$$

4.15 Velocity and acceleration

In normal conversation, we use 'speed' to describe how fast something is moving. However, when something is moving an important consideration, apart from its 'speed', is the direction in which it is travelling. When we take into account the direction in which an object or a body is moving be we use a more precise word, *velocity*. Velocity therefore means 'speed in a given direction'. For example, an object moving directly from point A to point B might have a velocity of 10 m/s. The same object moving at the same speed but from B to A would have a velocity of −10 m/s. Notice that we have *arbitrarily* defined a positive velocity as speed in the direction A to B.

Velocity is defined as the ratio of distance travelled to the time taken. Hence:

$$v = \frac{s}{t}$$

where v is velocity (in m/s), s is the distance (in m) and t is the time (in s).

Example 4.34

A cruise missile travels in a straight line at a constant speed of 295 m/s. How long will it take to reach a target which is 650 km away?

Re-arranging the formula $v = s/t$ to make t the subject gives:

$$t = \frac{s}{v}$$

From which:

$$t = \frac{650}{295} = \mathbf{2.203\,h}$$

If an object is moving at a constant velocity (i.e. its velocity is neither increasing nor decreasing) then its acceleration is zero. If the velocity of the object is increasing the object is undergoing *acceleration*. Conversely, if the velocity of the object is decreasing the object is undergoing *deceleration*.

Acceleration is defined as the ratio of change in velocity to the time. Hence:

$$a = \frac{v - u}{t}$$

where a is acceleration (in m/s^2), u is the velocity (in m/s) at the beginning of the time interval, v is the velocity (in m/s) at the end of the time interval, and t is the time (in s).

Note that, when v is greater than u, a will take a positive sign (acceleration). Conversely, when u is greater than v, a will have a negative sign (deceleration).

Example 4.35

An aircraft on a taxiway accelerates from 19 to 27 m/s in 10 s. What acceleration does the aircraft experience?

In this example, $u = 19$ m/s, $v = 27$ m/s and $t = 10$ s. Substituting these values into the equations for acceleration gives:

$$a = \frac{v - u}{t} = \frac{27 - 18}{10} = \frac{9}{10} = \textbf{0.9 m/s}^2$$

Example 4.36

The braking system fitted to a high-speed train is able to produce a constant deceleration 7.5 m/s^2. How long will it take for the train to come to rest if it is travelling at 120 m/s?

In this example, $u = 120$ m/s, $v = 0$ m/s (because the train *comes to rest*) and $a = -5.5$ m/s^2.

First we need to re-arrange the equation for acceleration in order to make t the subject:

$$t = \frac{v - u}{a} = \frac{0 - 120}{-5.5} = \frac{-120}{-5.5} = \textbf{21.82 s}$$

We can re-arrange the relationship that we met earlier to make the final velocity, v, the subject as follows:

$$a = \frac{v - u}{t}$$

Multiplying both sides by t gives:

$$a \times t = \frac{v - u}{t} \times t$$

or

$$at = v - u$$

Adding u to both sides gives:

$$at + u = v$$

Test your knowledge 4.29

1. A train is travelling at 24 m/s. What deceleration is required in order to bring the train to rest in a time of 30 s.
2. A car starts from rest and accelerates at 2.5 m/s^2 for 11 s. What is its final velocity?

Key point

Velocity means 'speed in a given direction'. Thus, unlike speed, velocity takes a positive or negative sign according to whether the motion is away from or towards a particular reference point.

Key point

When the velocity of a body is increasing it is said to be undergoing acceleration. When the velocity of a body is decreasing it is said to be undergoing deceleration.

or

$$v = u + at$$

Example 4.37

An overhead crane is travelling along a track at a velocity of 0.8 m/s. If the crane is then given a constant acceleration of 0.15 m/s^2, determine the velocity of the crane after 5 s.

Using $v = u + at$ we have:

$$v = u + at = 0.8 + (0.15 \times 5) = 0.8 + 0.75 = \mathbf{1.55\,m/s}$$

4.16 Graphs

Graphs provide us with a visual way of representing data. They can also be used to show, in a simple pictorial way, how one variable affects another variable. Many different types of graph are used in engineering. We shall start by looking at the most basic type, the straight line graph.

4.16.1 Straight line graphs

Earlier in Example 4.16 we introduced the idea of *proportionality*. In particular, we showed that the current flowing in a circuit was directly proportional to the voltage applied to it (from Ohm's law). We expressed this using the following mathematical notation:

$$I \propto V$$

where I represents the current and V represents output voltage.

We can illustrate this relationship using a simple graph showing current, I, plotted against voltage, V. Let us assume that the voltage applied to the circuit is varied in 1 V steps over the range 1–6 V and the circuit has a resistance of 3 Ω. By taking a set of measurements of V and I (see Figure 4.27) we would obtain the following date of corresponding values shown below:

Voltage, V (V)	1	2	3	4	5	6
Current, I (A)	0.33	0.66	1.0	1.33	1.66	2.0

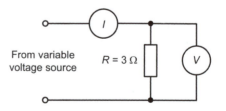

Figure 4.27 *Circuit showing experimental set-up to obtain readings of current and voltage*

The resulting graph is shown in Figure 4.28. To obtain the graph, a point is plotted for each pair of corresponding values for *V* and *I*. When all the points have been drawn they are connected together by drawing a line. Notice that, in this case, the line that connects the points together takes the form of a straight line. This is *always* the case when two variables are directly proportional to one another.

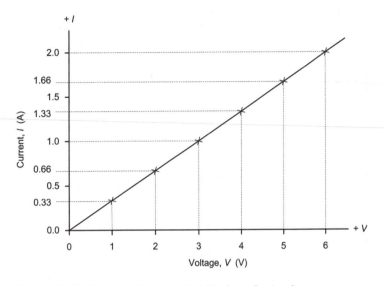

Figure 4.28 *Graph of current plotted against voltage*

It is conventional to show the *dependent variable* (in this case it is current, *I*) plotted on the vertical axis and the *independent variable* (in this case it is voltage, *V*) plotted on the horizontal axis. If you find these terms a little confusing, just remember that, what you now is usually plotted on the horizontal scale whilst what you do not know (and may be trying to find) is usually plotted on the vertical scale. In fact, the graph contains the same information regardless of which way round it is drawn!

Example 4.38

The following measurements are made on an electronic component:

Temperature, θ (°C)	10	20	30	40	50	60
Resistance, R (Ω)	105	110	115	120	125	130

Plot the graph showing how resistance varies with temperature. Determine the resistance of the component at 0°C and suggest the relationship that exists between resistance and temperature.

The results of the experiment are shown plotted in graphical form in Figure 4.29. Note that the graph consists of a straight line but that it does not pass through the *origin* of the graph (i.e. the point at which θ and R are 0°C and 0 Ω, respectively). The second most important feature to note (after having noticed that the graph is a straight line) is that, when $\theta = 0$°C, $R = 100\,\Omega$.

The following data was obtained whilst making measurements on an overhead conveyor:

Time, t (s)	Velocity, v (m/s)
0	0.5
1	1.1
2	1.7
3	2.3
4	2.9
5	3.5
6	4.1

Plot a graph showing how velocity varies with time. Use the graph to determine the value of velocity when $t = 2.5$ s and also determine the value of acceleration.

Key point

Graphs provide us with a way of visualising data. When a graph is plotted the independent variable is usually plotted on the horizontal axis (the x-axis) whilst the dependent variable is usually plotted on the vertical axis (y-axis).

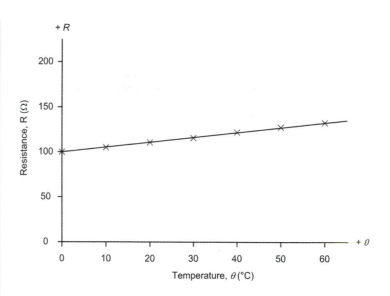

Figure 4.29 *Graph of resistance plotted against temperature*

By looking at the graph we could suggest a relationship (i.e. an *equation*) that will allow us to find the resistance, R, of the component at any given temperature, θ. The relationship is simply:

$$R = 100 + \frac{\theta}{2}\,\Omega$$

If you need to check that this works, just try inserting a few pairs of values from those given in the table. You should find that the equation balances every time!

4.16.2 Straight line graphs

The shape of a graph is dictated by the equation that connects its two variables. For example, the general equation for a straight line (like those shown earlier) takes the form:

$$y = mx + c$$

where y is the *dependent variable* (plotted on the vertical or y-axis), x is the *independent variable* (plotted on the horizontal or x-axis), m is the slope (or *gradient*) of the graph and c is the *intercept* on the y-axis. Figure 4.30 shows this information plotted on a graph.

The values of m (the gradient) and c (the y-axis intercept) are useful when quoting the specifications for electronic components. In the case of Example 4.38, the electronic component being tested (in this case a *thermistor*) has:

- a resistance of $100\,\Omega$ at 0°C (thus $c = 100\,\Omega$)
- a characteristic that exhibits an increase in resistance of $0.5\,\Omega$ per °C (thus $m = 0.5\,\Omega/°C$).

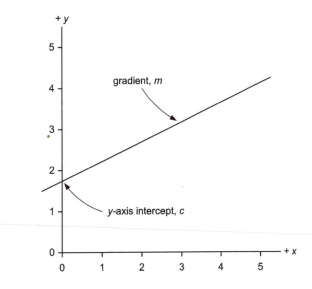

Figure 4.30 *Straight line law*

4.16.3 Square law graphs

Of course, not all graphs have a straight line shape. In the previous example we saw a graph that, whilst substantially linear, became distinctly curved at one end. Many graphs are curved rather than linear. One common type of curve is the square law. To put this into context, consider the relationship between the power developed in a load resistor and the current applied to it. Assuming that the load has a resistance of 15 Ω we could easily construct data showing corresponding values of power and current, as shown below:

Current, I (A)	0.25	0.5	0.75	1	1.25	1.5
Power, P (W)	0.94	3.75	8.44	15	23.44	33.75

We can plot this information on a graph showing power, P, on the vertical axis plotted against current, I, on the horizontal axis. In this case, P is the *dependent variable* and I is the *independent variable*. The graph is shown in Figure 4.31.

It can be seen that the relationship between P and I in Figure 4.31 is far from linear. The relationship is, in fact, a *square law relationship*. We can actually deduce this from what we know about the power dissipated in a circuit and the current flowing in the circuit. You may recall that:

$P = I^2 R$

where P represents power in watts, I is current in amps, and R is resistance in ohms.

Since R remains constant, we can deduce that:

$P \propto I^2$

In words, we would say that 'power is proportional to current squared'. Many other examples of square law relationships are found in engineering.

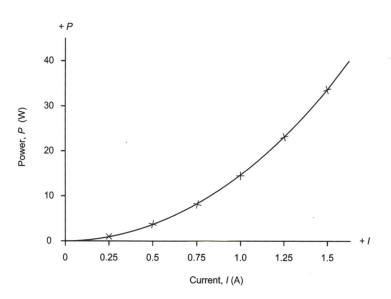

Figure 4.31 *Power plotted against current*

4.16.4 More complex graphs

Many more complex graphs exist and Figure 4.32 shows some of the most common types. Note that these graphs have all been plotted over the range $x = \pm 4$, $y = \pm 4$. Each graph consists of four quadrants. These are defined as follows (see Figure 4.33):

First quadrant	Values of x and y are both positive
Second quadrant	Values of x are negative whilst those for y are positive
Third quadrant	Values of x and y are both negative
Fourth quadrant	Values of x are positive whilst those for y are negative.

The straight line relationship, $y = x$, is shown in Figure 4.32(a). This graph consists of a straight line with a gradient of 1 that passes through the *origin* (i.e. the point where $x = 0$ and $y = 0$). The graph has values in the first and third quadrants.

The relationship $y = x^2$ is shown in Figure 4.32(b). This graph also passes through the origin but its gradient changes, becoming steeper for larger values of x. As you can see, the graph has values in the first and second quadrants.

The graph of $y = x^3$ is shown in Figure 4.32(c). This *cubic law* graph is steeper than the square law of Figure 4.32(b) and it has values in the first and third quadrants.

Figure 4.32(d) shows the graph of $y = x^4$. This graph is even steeper than those in Figure 4.32(b) and (c). Like the square law graph of Figure 4.32(b), this graph has values in the first and second quadrants.

The graph of $y = x^5$ is shown in Figure 4.32(e). Like the cubic law graph of Figure 4.32(c), this graph has values in the first and third quadrants.

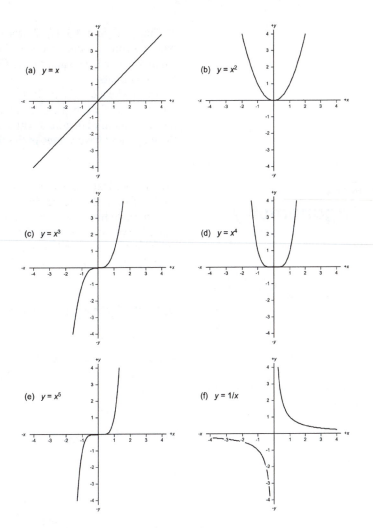

Figure 4.32 *Some complex graphs*

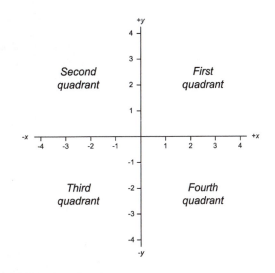

Figure 4.33 *Quadrants*

Finally, Figure 4.32(f) shows the graph of $y = 1/x$ (or $y = x^{-1}$). Note how the y values are very large for small values of x and very small for very large values of x. This graph has values in the first and third quadrants.

If you take a careful look at Figure 4.32 you should notice that, for odd powers of x (i.e. x^1, x^3, x^5 and x^{-1}) the graph will have values in the first and third quadrant whilst for even powers of x (i.e. x^2 and x^4) the graph will have values in the first and second quadrants.

4.17 Trigonometry

Trigonometrical ratios are to do with the way in which we measure angles. Take a look at the right-angled triangle shown in Figure 4.34. This triangle has three sides; a, b, and c. The angle that we are interested in (we have used the Greek symbol θ to denote this angle) is adjacent to side a and is opposite to side b. The third side of the triangle (the *hypotenuse*) is the longest side of the triangle.

4.17.1 Pythagoras' theorem

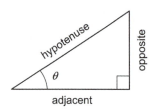

Figure 4.34 *A right-angled triangle*

In Figure 4.35, the theorem of Pythagoras states that, for a right-angled triangle, 'the square on the hypotenuse is equal to the sum of the squares on the other two sides'. Writing this as an equation we arrive at:

$$c^2 = a^2 + b^2$$

where c is the hypotenuse and a and b are the other two sides.

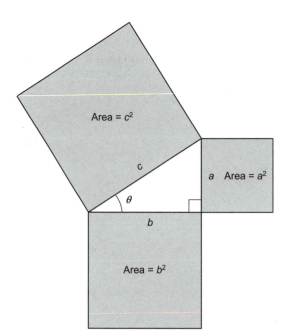

Figure 4.35 *Pythagoras' theorem*

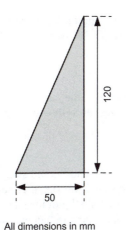

All dimensions in mm

Figure 4.36 *See Example 4.39*

Taking square roots of both sides of the equation we can see that:

$$c = \sqrt{a^2 + b^2}$$

Thus if we know two of the sides (e.g. a and b) of a right-angled triangle we can easily find the third side (c).

Example 4.39

A triangular brace made from sheet metal is shown in Figure 4.36. Determine the length of the third side.

From Figure 4.36 the two perpendicular sides of the triangle have lengths of 50 and 120 mm. The remaining side can be calculated from:

$$c = \sqrt{50^2 + 120^2} = \sqrt{2500 + 14{,}400} = \sqrt{16{,}900} = \textbf{130 mm}$$

4.17.2 Trigonometrical ratios

The ratios a/c, b/c and a/b are known as the basic *trigonometric ratios*. They are known as sine (*sin*), cosine (*cos*) and tangent (*tan*) of angle θ, respectively (see Figure 4.34). Thus:

$$\sin \theta = \frac{\text{opposite}}{\text{hypotenuse}} = \frac{a}{c}$$

and

$$\cos \theta = \frac{\text{adjacent}}{\text{hypotenuse}} = \frac{b}{c}$$

and

$$\tan \theta = \frac{\text{opposite}}{\text{adjacent}} = \frac{a}{b}$$

4.17.3 Trigonometrical equations

Equations that involve trigonometrical expressions are known as trigonometrical equations. Fortunately they are not quite so difficult to understand as they sound! Consider the equation:

$$\sin \theta = 0.5$$

This equation can be solved quite easily using a calculator. However, before doing so, you need to be sure to select the correct mode for expressing angles on your calculator. If you are using a 'scientific calculator' you will find that you can set the angular mode to either *radian* measure or *degrees*. A little later we will explain the difference between these two angular measures but for the time being we shall just use degrees.

Key point

Pythagoras' theorem states that, for a right-angled triangle, the square on the hypotenuse is equal to the sum of the squares on the other two sides. Thus, if we know the lengths of two sides of a right-angled triangle we can always find the length of the third side.

If you solve the equation (by keying in 0.5 and pressing the *inverse sine* function keys) you should see the result 30° displayed on your calculator. Hence we can conclude that:

$\sin 30° = 0.5$

Actually, a number of other angles will give the same result! Try pressing the sine function key and entering the following angles in turn:

30°, 210°, 390° and 570°

They should all produce the same result, 0.5! This should suggest to you that the graph of the sine function repeats itself (i.e. the shape of the graph is *periodic*). In the next section we shall plot the sine function but, before we do we shall take a look at using radian measure to specify angles.

4.17.4 Radians

The *radian* is defined as the angle subtended by an arc of a circle equal in length to the radius of the circle. This relationship is illustrated in Figure 4.37.

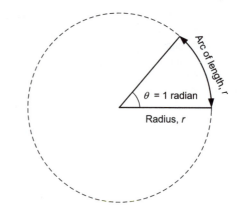

Figure 4.37 *Definition of the radian*

The circumference, *l*, of a circle is related to its radius, *r*, according to the formula:

$l = 2\pi r$

Thus,

$$r = \frac{l}{2\pi}$$

Now, since there are 360° in one complete revolution we can deduce that 1 radian is the same as $360°/2\pi = 57.3°$. On other words, to convert:

- degrees to radians divide by 57.3
- radians to degrees multiply by 57.3

It is important to note that one complete cycle of a periodic function (i.e. a waveform) occurs in a time, T. This is known as the *periodic time* or just the *period*. In a time interval equal to T, the angle will have changed by 360°. The relationship between time and angle expressed in degrees is thus:

$$\theta = \frac{T}{t} \times 360° \text{ and } t = \frac{T}{\theta} \times 360°$$

Thus, if one complete cycle (360°) is completed in 0.02 s (i.e. $T = 20$ ms) an angle of 180° will correspond to a time of 0.01 s (i.e. $t = 10$ ms).

Conversely, if we wish to express angles in radians:

$$\theta = \frac{T}{t} \times 2\pi \text{ and } t = \frac{T}{\theta} \times 2\pi$$

Thus, if one complete cycle (2π radians) is completed in 0.02 s (i.e. $T = 20$ ms) an angle of π radians will correspond to a time of 0.01 (i.e. $t = 10$ ms).

4.17.5 Graphs of trigonometrical functions

To plot a graph of $y = \sin \theta$ we can construct a table of values of $\sin \theta$ as θ is varied from 0° to 360° in suitable steps (say, every 30°). This exercise (carried out using a scientific calculator) will produce data that looks something like this:

Angle, θ	0°	30°	60°	90°	120°	150°	180°
$\sin \theta$	0	0.5	0.866	1	0.866	0.5	0

Angle, θ	210°	240°	270°	300°	330°	360°	
$\sin \theta$	−0.5	−0.866	−1	−0.866	−0.5	0	

Plotting the values given above reveals the graph shown in Figure 4.38.

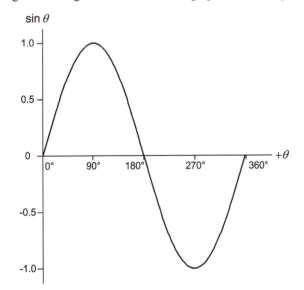

Figure 4.38 *Graph of y = sin θ*

The following data was obtained from measurements made on a crank arm:

Angle, θ	Distance, d (m)
0°	0
30°	1.1
60°	1.9
90°	2.2
120°	1.9
150°	1.1
180°	0
210°	−1.1
240°	−1.9
270°	−2.2
300°	−1.9
330°	−1.1
360°	0

Plot a graph showing how distance varies with angle and use it to find the values of θ that correspond to a distance of 1 m. Label your axes clearly. Also determine the maximum positive and negative displacements of the crank.

We can use the same technique to produce graphs of $\cos\theta$ and $\tan\theta$, as shown in Figures 4.39 and 4.40.

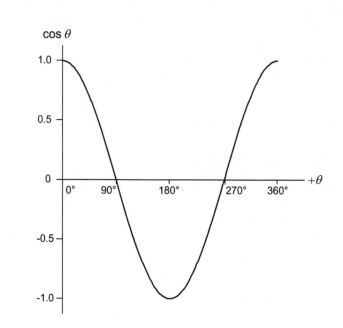

Figure 4.39 *Graph of* $y = \cos\theta$

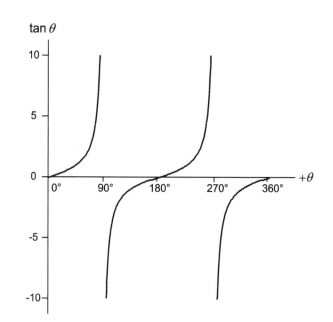

Figure 4.40 *Graph of* $y = \tan\theta$

Answers to numerical test your knowledge questions

Test your knowledge 4.1

(a) -11
(b) $+12$
(c) -45
(d) $+77$
(e) -5

Test your knowledge 4.2

Only statement (a) is correct

Test your knowledge 4.3

1. Hertz
2. Coulomb
3. Newton
4. m/s/s/ or m/s^2 or m s^{-2}
5. I
6. 40 W
7. 10 m/s^2
8. 150 C

Test your knowledge 4.4

1. 0.05 V
2. 1250 mm
3. 9740 MHz
4. 0.44 kV
5. 75 μm
6. 15,620 kHz
7. 0.57 mA
8. 220 nF
9. 0.47 MΩ
10. 0.254 mm

Test your knowledge 4.5

1. 13.5 tonnes
2. 74.568 mph
3. 5,455.32 l
4. 193,054.4 Pa

Test your knowledge 4.6

(a) 10.049
(b) 2.222
(c) 6.0119
(d) 751.07

(e) 0.5778
(f) 2.2635
(g) 35.426
(h) 9.256
(i) 7.29
(j) 1.4681

Test your knowledge 4.7

Correct statements are (b), (c), (f), (g) and (i).

Test your knowledge 4.8

1. 9.8
2. 1.18
3. 26.67
4. 8200
5. 41.58

Test your knowledge 4.9

1. 25
2. 25.94 (nearly 26 laps)

Test your knowledge 4.10

45 psi

Test your knowledge 4.11

1. $\rho = \dfrac{M}{V}$

2. 0.1292

Test your knowledge 4.12

All of the expressions are true

Test your knowledge 4.13

1. $1.66 \times 10^5\,\text{N}$
2. $5.15 \times 10^3\,\text{m/s}$
3. $3.77 \times 10^{-9}\,\text{F}$
4. $5.2 \times 10^{-7}\,\text{A}$
5. $4.75 \times 10^{-2}\,\text{W}$
6. $2.2 \times 10^4\,\Omega$

Test your knowledge 4.14

1. $2.65 \times 10^3 \, \text{N}$
2. $525 \times 10^{-3} \, \text{V}$
3. $22 \times 10^{-3} \, \Omega$
4. $65 \times 10^3 \, \text{m}$
5. $825.5 \times 10^3 \, \text{Hz}$
6. $6.5 \times 10^{-3} \, \text{A}$

Test your knowledge 4.15

1. 2.65 kN
2. 525 mV
3. 22 mΩ
4. 65 km
5. 825.5 kHz
6. 6.5 mA

Test your knowledge 4.16

1. (a) 1.9950E4   (b) 7.5E$-$3
2. (a) 1.59125×10^2 (b) 1.915×10^{-6}

Test your knowledge 4.17

1. 11.1 kPa
2. $3.68 \, \text{m/s}^2$

Test your knowledge 4.18

(a) $8 \, \text{m}^2$
(b) $5 \, \text{m}^2$
(c) $7 \, \text{m}^2$
(d) $2.625 \, \text{m}^2$
(e) $1525 \, \text{m}^2$

Test your knowledge 4.19

$A = 45°, \quad B = 60°, \quad C = 75°$

Test your knowledge 4.20

(a) $2 \, \text{m}^2$
(b) $2.5 \, \text{m}^2$
(c) $2175 \, \text{mm}^2$

Test your knowledge 4.21

Total area $= 50.4\,\text{m}^2$

Test your knowledge 4.22

1. $29.845\,\text{mm}^2$
2. $34,558\,\text{km}$

Test your knowledge 4.23

$1.672\,\text{m}^3$

Test your knowledge 4.24

$84,960\,\text{mm}^3$

Test your knowledge 4.25

$8.4823\,\text{m}^3$

Test your knowledge 4.26

1. $4.66\,\text{kN}$
2. $186.3\,\text{kg}$
3. $39.24\,\text{kPa}$

Test your knowledge 4.27

1. $1.875\,\text{kN}$
2. $4.29\,\text{m/s}^2$

Test your knowledge 4.28

$184.6\,\text{mph}$

Test your knowledge 4.29

1. $0.8\,\text{m/s}^2$
2. $27.5\,\text{m/s}$

Test your knowledge 4.30

$2\,\text{m/s}$, $0.6\,\text{m/s}^2$

Test your knowledge 4.31

(i) 225 m/s
(ii) 362.5 m/s
(iii) 3.182 s

Test your knowledge 4.32

(a) 3.78 m
(b) 5.66 s

Test your knowledge 4.33

1. (a) 0.96   (b) 3.054
2. (a) 11.46°   (b) 143.25°
3. (a) 0.966   (b) −0.766   (c) 0.364

Test your knowledge 4.34

27° and 153°; ±2.2 m

Appendix:
Engineering data

Abbreviations

2D	two-dimensional	FM	frequency modulation
3D	three-dimensional	FMS	flexible manufacturing system
a.c.	alternating current	HSE	Health and Safety Executive
ACW	anticlockwise	I/O	input/output
ADC	analogue-to-digital converter	JIT	just-in-time
AF	audio frequency	KE	kinetic energy
ALU	arithmetic and logic unit	LCD	liquid crystal display
AM	amplitude modulation	LDR	light dependent resistor
ASCII	American standard code for	LED	light emitting diode
	information interchange	LHS	left-hand side
BCD	binary coded decimal	LSB	least-significant bit
BJT	bipolar junction transistor	LVDT	linear variable differential
BM	bending moment		transformer
BS	British Standards	MOS	metal oxide semiconductor
CAD	computer-aided design	MSB	most-significant bit
CAE	computer-aided engineering	MTBF	mean time between failure
CAM	computer-aided manufacture	PC	personal computer
CIMS	computer-integrated manufacturing	PCB	printed circuit board
	system	PDS	product design specification
CNC	computer numerical controlled	PE	potential energy
COSHH	control of substance hazardous	PEEK	polyetheretherketone
	to health	PERT	programme evaluation and review
CPM	critical path method		technique
CPU	central processing unit	PLA	programmed logic array
CRT	cathode ray tube	PLC	programmable logic controllers
CW	clockwise	PLD	programmed logic device
DAC	digital-to-analogue converter	PROM	programmable read-only memory
d.c.	direct current	PTFE	polytetraflouroethylene
DFM	digital frequency meter	PVC	polyvinylchloride
DPM	digital panel meter	PWM	pulse width modulated
e.m.f.	electromotive force	QA	quality assurance
FEA	finite element analysis	QC	quality control
FEM	finite element model	R&D	research and development
FET	field effect transistor	RAM	random access memory

RF	radio frequency	SNR	signal-to-noise ratio
RHS	right-hand side	TQ	total quality
RMS	root mean square	TQM	total quality management
ROM	read-only memory	UDL	uniformly distributed load
rpm	revolutions per minute	VA	volt-amperes
SHM	simple harmonic motion	VDU	visual display unit
S/N	signal-to-noise	WD	work done

Conversion table: mm to inches

mm	inch	mm	inch
0.0	0.0000	19.0	0.7480
0.5	0.0197	19.5	0.7677
1.0	0.0394	20.0	0.7874
1.5	0.0591	20.5	0.8071
2.0	0.0787	21.0	0.8268
2.5	0.0984	21.5	0.8465
3.0	0.1181	22.0	0.8661
3.5	0.1378	22.5	0.8858
4.0	0.1575	23.0	0.9055
4.5	0.1772	23.5	0.9252
5.0	0.1969	24.0	0.9449
5.5	0.2165	24.5	0.9646
6.0	0.2362	25.0	0.9843
6.5	0.2559	25.5	1.0039
7.0	0.2756	26.0	1.0236
7.5	0.2953	26.5	1.0433
8.0	0.3150	27.0	1.0630
8.5	0.3346	27.5	1.0827
9.0	0.3543	28.0	1.1024
9.5	0.3740	28.5	1.1220
10.0	0.3937	29.0	1.1417
10.5	0.4134	29.5	1.1614
11.0	0.4331	30.0	1.1811
11.5	0.4528	30.5	1.2008
12.0	0.4724	31.0	1.2205
12.5	0.4921	31.5	1.2402
13.0	0.5118	32.0	1.2598
13.5	0.5315	32.5	1.2795
14.0	0.5512	33.0	1.2992
14.5	0.5709	33.5	1.3189
15.0	0.5906	34.0	1.3386
15.5	0.6102	34.5	1.3583
16.0	0.6299	35.0	1.3780
16.5	0.6496	35.5	1.3976
17.0	0.6693	36.0	1.4173
17.5	0.6890	36.5	1.4370
18.0	0.7087	37.0	1.4567
18.5	0.7283	37.5	1.4764

Conversion table: mm to inches (continued)

mm	inch
38.0	1.4961
38.5	1.5157
39.0	1.5354
39.5	1.5551
40.0	1.5748
40.5	1.5945
41.0	1.6142
41.5	1.6339
42.0	1.6535
42.5	1.6732
43.0	1.6929
43.5	1.7126
44.0	1.7323

mm	inch
44.5	1.7520
45.0	1.7717
45.5	1.7913
46.0	1.8110
46.5	1.8307
47.0	1.8504
47.5	1.8701
48.0	1.8898
48.5	1.9094
49.0	1.9291
49.5	1.9488
50.0	1.9685

Conversion table: inches to mm

inch	mm
0	0.00
0.1	2.54
0.2	5.08
0.3	7.62
0.4	10.16
0.5	12.70
0.6	15.24
0.7	17.78
0.8	20.32
0.9	22.86
1.0	25.40
1.1	27.94
1.2	30.48
1.3	33.02
1.4	35.56
1.5	38.10
1.6	40.64
1.7	43.18
1.8	45.72
1.9	48.26
2.0	50.80
2.1	53.34
2.2	55.88
2.3	58.42
2.4	60.96
2.5	63.50
2.6	66.04
2.7	68.58

inch	mm
2.8	71.12
2.9	73.66
3.0	76.20
3.1	78.74
3.2	81.28
3.3	83.82
3.4	86.36
3.5	88.90
3.6	91.44
3.7	93.98
3.8	96.52
3.9	99.06
4.0	101.60
4.1	104.14
4.2	106.68
4.3	109.22
4.4	111.76
4.5	114.30
4.6	116.84
4.7	119.38
4.8	121.92
4.9	124.46
5.0	127.00
5.1	129.54
5.2	132.08
5.3	134.62
5.4	137.16
5.5	139.70

Conversion table: inches to mm (continued)

inch	mm	inch	mm
5.6	142.24	7.9	200.66
5.7	144.78	8.0	203.20
5.8	147.32	8.1	205.74
5.9	149.86	8.2	208.28
6.0	152.40	8.3	210.82
6.1	154.94	8.4	213.36
6.2	157.48	8.5	215.90
6.3	160.02	8.6	218.44
6.4	162.56	8.7	220.98
6.5	165.10	8.8	223.52
6.6	167.64	8.9	226.06
6.7	170.18	9.0	228.60
6.8	172.72	9.1	231.14
6.9	175.26	9.2	233.68
7.0	177.80	9.3	236.22
7.1	180.34	9.4	238.76
7.2	182.88	9.5	241.30
7.3	185.42	9.6	243.84
7.4	187.96	9.7	246.38
7.5	190.50	9.8	248.92
7.6	193.04	9.9	251.46
7.7	195.58	10.0	254.00
7.8	198.12		

Conversion table: fractions of an inch to decimal and mm

inch		mm	inch		mm
Fraction	Decimal		Fraction	Decimal	
0	0.00000	0.00000	15/32	0.46875	11.90625
1/32	0.03125	0.79375	1/2	0.50000	12.70000
1/16	0.06250	1.58750	17/32	0.53125	13.49375
3/32	0.09375	2.38125	9/16	0.56250	14.28750
1/8	0.12500	3.17500	19/32	0.59375	15.08125
5/32	0.15625	3.96875	5/8	0.62500	15.87500
3/16	0.18750	4.76250	21/32	0.65625	16.66875
7/32	0.21875	5.55625	11/16	0.68750	17.46250
1/4	0.25000	6.35000	23/32	0.71875	18.25625
9/32	0.28125	7.14375	3/4	0.75000	19.05000
5/16	0.31250	7.93750	25/32	0.78125	19.84375
11/32	0.34375	8.73125	13/16	0.81250	20.63750
3/8	0.37500	9.52500	27/32	0.84375	21.43125
13/32	0.40625	10.31875	7/8	0.87500	22.22500
7/16	0.43750	11.11250	29/32	0.90625	23.01875

Conversion table: fractions of an inch to decimal and mm (continued)

inch		mm	inch		mm
Fraction	Decimal		Fraction	Decimal	
15/16	0.93750	23.81250	2⅜	2.37500	60.32500
31/32	0.96875	24.60625	2 13/32	2.40625	61.11875
1	1.00000	25.40000	2 7/16	2.43750	61.91250
1 1/32	1.03125	26.19375	2 15/32	2.46875	62.70625
1 1/16	1.06250	26.98750	2½	2.50000	63.50000
1 3/32	1.09375	27.78125	2 17/32	2.53125	64.29375
1⅛	1.12500	28.57500	2 9/16	2.56250	65.08750
1 5/32	1.15625	29.36875	2 19/32	2.59375	65.88125
1 3/16	1.18750	30.16250	2⅝	2.62500	66.67500
1 7/32	1.21875	30.95625	2 21/32	2.65625	67.46875
1¼	1.25000	31.75000	2 11/16	2.68750	68.26250
1 9/32	1.28125	32.54375	2 23/32	2.71875	69.05625
1 5/16	1.31250	33.33750	2¾	2.75000	69.85000
1 11/32	1.34375	34.13125	2 25/32	2.78125	70.64375
1⅜	1.37500	34.92500	2 13/16	2.81250	71.43750
1 13/32	1.40625	35.71875	2 27/32	2.84375	72.23125
1 7/16	1.43750	36.51250	2⅞	2.87500	73.02500
1 15/32	1.46875	37.30625	2 29/32	2.90625	73.81875
1½	1.50000	38.10000	2 15/16	2.93750	74.61250
1 17/32	1.53125	38.89375	2 31/32	2.96875	75.40625
1 9/16	1.56250	39.68750	3	3.00000	76.20000
1 19/32	1.59375	40.48125	3 1/32	3.03125	76.99375
1⅝	1.62500	41.27500	3 1/16	3.06250	77.78750
1 21/32	1.65625	42.06875	3 3/32	3.09375	78.58125
1 11/16	1.68750	42.86250	3⅛	3.12500	79.37500
1 23/32	1.71875	43.65625	3 5/32	3.15625	80.16875
1¾	1.75000	44.45000	3 3/16	3.18750	80.96250
1 25/32	1.78125	45.24375	3 7/32	3.21875	81.75625
1 13/16	1.81250	46.03750	3¼	3.25000	82.55000
1 27/32	1.84375	46.83125	3 9/32	3.28125	83.34375
1⅞	1.87500	47.62500	3 5/16	3.31250	84.13750
1 29/32	1.90625	48.41875	3 11/32	3.34375	84.93125
1 15/16	1.93750	49.21250	3⅜	3.37500	85.72500
1 31/32	1.96875	50.00625	3 13/32	3.40625	86.51875
2	2.00000	50.80000	3 7/16	3.43750	87.31250
2 1/32	2.03125	51.59375	3 15/32	3.46875	88.10625
2 1/16	2.06250	52.38750	3½	3.50000	88.90000
2 3/32	2.09375	53.18125	3 17/32	3.53125	89.69375
2⅛	2.12500	53.97500	3 9/16	3.56250	90.48750
2 5/32	2.15625	54.76875	3 19/32	3.59375	91.28125
2 3/16	2.18750	55.56250	3⅝	3.62500	92.07500
2 7/32	2.21875	56.35625	3 21/32	3.65625	92.86875
2¼	2.25000	57.15000	3 11/16	3.68750	93.66250
2 9/32	2.28125	57.94375	3 23/32	3.71875	94.45625
2 5/16	2.31250	58.73750	3¾	3.75000	95.25000
2 11/32	2.34375	59.53125	3 25/32	3.78125	96.04375

Conversion table: fractions of an inch to decimal and mm (continued)

inch		mm		inch		mm
Fraction	*Decimal*			*Fraction*	*Decimal*	
$3\frac{13}{16}$	3.81250	96.83750		$4\frac{7}{16}$	4.43750	112.71250
$3\frac{27}{32}$	3.84375	97.63125		$4\frac{15}{32}$	4.46875	113.50625
$3\frac{7}{8}$	3.87500	98.42500		$4\frac{1}{2}$	4.50000	114.30000
$3\frac{29}{32}$	3.90625	99.21875		$4\frac{17}{32}$	4.53125	115.09375
$3\frac{15}{16}$	3.93750	100.01250		$4\frac{9}{16}$	4.56250	115.88750
$3\frac{31}{32}$	3.96875	100.80625		$4\frac{19}{32}$	4.59375	116.68125
4	4.00000	101.60000		$4\frac{5}{8}$	4.62500	117.47500
$4\frac{1}{32}$	4.03125	102.39375		$4\frac{21}{32}$	4.65625	118.26875
$4\frac{1}{16}$	4.06250	103.18750		$4\frac{11}{16}$	4.68750	119.06250
$4\frac{3}{32}$	4.09375	103.98125		$4\frac{23}{32}$	4.71875	119.85625
$4\frac{1}{8}$	4.12500	104.77500		$4\frac{3}{4}$	4.75000	120.65000
$4\frac{5}{32}$	4.15625	105.56875		$4\frac{25}{32}$	4.78125	121.44375
$4\frac{3}{16}$	4.18750	106.36250		$4\frac{13}{16}$	4.81250	122.23750
$4\frac{7}{32}$	4.21875	107.15625		$4\frac{27}{32}$	4.84375	123.03125
$4\frac{1}{4}$	4.25000	107.95000		$4\frac{7}{8}$	4.87500	123.82500
$4\frac{9}{32}$	4.28125	108.74375		$4\frac{29}{32}$	4.90625	124.61875
$4\frac{5}{16}$	4.31250	109.53750		$4\frac{15}{16}$	4.93750	125.41250
$4\frac{11}{32}$	4.34375	110.33125		$4\frac{31}{32}$	4.96875	126.20625
$4\frac{3}{8}$	4.37500	111.12500		5	5.00000	127.00000
$4\frac{13}{32}$	4.40625	111.91875				

Data on selected materials

Material	Density (mg/m^3)	Young's modulus (GPa)	Strength (MPa)	Ductility (%)	Toughness, K_{IC} $(MPa\, m^{1/2})$	Specific modulus (GPa)/ (mg/m^3)	Specific strength (MPa)/ (mg/m^3)
Ceramics							
Alumina	3.87	382	332	0	4.9	99	86
Magnesia	3.60	207	230	0	1.2	58	64
Silicon nitride		166	210	0	4.0		
Zirconia	5.92	170	900	0	8.6	29	152
β-Sialon	3.25	300	945	0	7.7	92	291
Metals							
Aluminium	2.70	69	77	47	~30	26	29
Aluminium alloy	2.83	72	325	18	~25–30	25	115
Brass	8.50	100	550	70	–	12	65
Nickel alloy	8.18	204	1200	26	~50–80	25	147
Steel mild	7.86	210	460	35	~50	27	59
Titanium alloy	4.56	112	792	20	~55–90	24	174
Polymers							
Epoxy	1.12	4	50	4	1.5	4	36
Nylon 6.6	1.14	2	70	60	3–4	18	61
Polyetheretherketone	1.30	4	70		1.0	3	54
Polymethylmethacrylate	1.19	3	50	3	1.5	3	42
Polystyrene	1.05	3	50	2	1.0	3	48
Polyvinylchloride (rigid)	1.70	3	60	15	4.0	2	35

'How to' Index

How to carry out a brainstorming session   12
How to construct a first-angle projection   62
How to construct a third-angle projection   65
How to construct an isometric drawing   58
How to construct an oblique drawing   56
How to draw a bar chart   10
How to draw a block diagram   29
How to draw a flow chart   30
How to draw a mind map   14
How to draw a pie chart   10
How to draw a schematic diagram   33
How to organise a design folder   90
How to prepare a design brief   5
How to prepare a production plan   103
How to prepare a specification   101
How to use an evaluation matrix   16
How to write a questionnaire   8
How to write a technical report   92

Index

2D drawing, 38
2D layout drawing, 40
2D sketch, 26
3D CAD drawing, 40
3D assembly diagram, 39
3D drawing, 242
3D model, 43
3D sketch, 26

ADC, 263
ALU, 254
Absolute dimensioning, 48
Absolute value, 274
Acceleration, 277, 307, 308
Accidents, 210
Acrylics, 122
Actuator, 70, 71
Address bus, 261, 262
Adherend, 193
Adhesion, 193, 195
Adhesive, 193
Admiralty brass, 116
Aerospace, 230
Aesthetics, 20
Alloys, 115
Aluminium, 118
Ammeter, 80
Amount of substance, 276
Ampere, 276
Amplifier, 83
Analogue signal, 247
Analogue system, 247
Analogue-to-digital converter, 263
Angle, 296
Angle grinder, 205
Angle plate, 171
Angular measure, 139, 296
Anode, 203
Approximation, 284

Arc welding, 189
Area, 294, 296, 297, 302
Arithmetic logic unit, 254, 261
Arrowheads, 45
Assembly, 35, 184
Assembly diagram, 39
AutoCAD, 39
AutoSketch, 38, 39
Automation, 255
Autopilot, 246
Auxiliary view, 55

BS 2197, 42
BS 308, 34, 42
BS 3939, 42
BS 4058, 30
BS 4500, 42, 50
BS PD 6112, 20
BS PP 7307, 42
Bakelite, 121
Bar chart, 8, 9, 10
Battery, 79
Bench, 153
Blacking, 207
Block diagram, 29
Bolt, 186, 187
Bolted connection, 199
Bolts, 125
Bonded joint, 194, 196
Bonding, 193
Boring, 178, 181
Bottoming, 164
Bought-in parts, 35
Brainstorming, 11, 12
Brass, 115, 116
Brazing, 193
British Standards, 41
Bronze alloys, 117
Bulletin board, 239

Bus, 262
Butt joint, 126, 194

CAD, 38, 233, 240, 241
CAD/CAM, 38
CAE, 38, 240, 243
CAM, 233, 240
CD, 266, 268
CD-E, 268
CD-R, 268
CD-ROM, 268
CERN, 236
CIM, 240, 243
CISC, 255
CNC, 38, 243, 261
CPU, 253, 261
Cabinet oblique projection, 56
Cable, 128
Calculator, 282
Caliper gauge, 143
Calipers, 137, 138
Candela, 276
Candidate solution, 16, 88
Capacitor, 79, 130, 132
Carbon resistor, 130
Carbon steel, 113, 114
Cast iron, 113
Casting, 183
Castle nut, 187
Cavalier oblique projection, 56
Cell, 79
Cellulose plastics, 122
Central processing unit, 253
Centre drill, 175
Centre lathe, 173, 174
Centre mark, 144, 145
Centre punch, 145, 146
Ceramics, 118, 119, 120
Chain dimensioning, 48
Chamfer, 181
Characteristics, 20
Charge, 277
Chart, 8, 9, 10, 30
Chassis, 67
Chemical blacking, 207
Chemical properties of materials, 107
Chemical treatment, 201
Chip, 83, 253, 254
Chipping, 157
Chisel, 156, 157, 158, 159
Chromium plating, 203
Chuck, 176, 177
Circle, 294, 297, 300, 301
Circuit diagram, 32, 79
Circuit schematic, 31
Circular coordinates, 150
Circumference, 301

Clamped connection, 199
Clearance, 186
Clearance angle, 155, 179
Client, 5, 19
Clock, 261
Clothing, 211, 212, 213
Coating, 207
Cohesion, 195
Coil, 132
Cold chisel, 156, 157
Collimating lens, 267
Colour code for resistors, 132
Compact disc, 266, 268
Complex instruction set computer, 255
Component parts, 98
Components, 106, 124, 127, 248
Composite materials, 251
Compound slide, 179
Computer aided design, 233, 240
Computer aided engineering, 240
Computer aided manufacture, 38, 233
Computer numerical control, 38, 243
Conditioning devices, 76
Conditions of use, 20
Conformance quality, 105
Conical surfaces, 178
Connectors, 129
Constants, 285
Construction lines, 61, 65
Consumer unit, 82
Control bus, 261, 262
Control program, 262
Control signals, 262
Control system, 245, 246, 255
Control technology, 245
Control unit, 261
Controlled process, 245
Controlled variable, 245
Controller, 245
Conveyor belt, 264
Coolant, 156
Coordinates, 150, 151
Copper, 115
Corrosion, 107
Cosine, 317, 319
Coulomb, 277
Counterboring, 172, 173
Countersinking, 172, 173
Cracking, 201
Crimped joint, 197
Crimping, 199
Crimping tool, 199
Criteria, 16
Cross-filing, 160
Cube, 302

Curing, 121, 250, 252
Current, 276
Cutting, 155, 156
Cutting angle, 159
Cutting edge, 170
Cutting plane, 52
Cutting tools, 154
Cylindrical components, 150

DAC, 263
DBMS, 234
DCV, 67, 68, 69, 70, 71, 74
DIN plug, 130
DIN socket, 130
DNA, 250
Data, 253
Data bus, 261, 262
Data link, 249
Database, 234, 235
Database management system, 234
Database manager, 234
Datum, 148, 151
Datum point, 151, 152
Decimal places, 284
Decimal point, 274
Definitions, 20
Degradation of materials, 108
Degrees, 317
Density, 306
Design brief, 5, 67, 87
Design criteria, 88
Design folder, 86, 88, 90
Design problem, 4
Design process, 2, 3
Design quality, 105
Design solution, 16, 85, 89
Design specification, 19, 88
DesignCAD, 38
Detail drawing, 37
Dezincification, 116
Diagram, 29, 30, 31, 32, 33, 66, 79
Diameter, 301
Die casting, 183
Die holder, 165
Digital signal, 247
Digital system, 247
Digital-to-analogue converter, 263
Dimensioned drawing 41
Dimensioning, 51
Dimensions, 48
Diode, 134
Diodes, 81
Dipping, 207, 208
Directional control valve, 67, 68
Distortion, 201
Dividers, 147
Dot punch, 146, 147
Draw-filing, 160

Drawing, 21, 24, 34, 35, 36, 37, 40,
    53, 111
Drawing conventions, 48
Drawing standards, 41
Drawing template, 34
Drawing zone, 34
Drift, 168
Drill, 167, 169, 170
Drill shank, 168
Drill tang, 169
Drilling, 166
Drilling machines, 166
Drunken thread, 165
Ductility, 110, 111

E-mail, 238
ELCB, 217
Earth leakage circuit breaker, 217
Economics, 21
Elasticity, 110
Electric charge, 277
Electric current, 276
Electric drill, 167, 216
Electric saw, 231
Electric shock, 215, 217
Electrical and electronic
    engineering, 230
Electrical diagram, 79
Electrical hazards, 212
Electrical properties of materials,
    108
Electrical safety, 214
Electrician's screwdriver, 98, 99
Electrolyte, 203
Electrolytic capacitor, 132
Electrolytic galvanizing, 204
Electromechanical devices, 97
Electronic calculator, 282
Electronic circuit, 32, 81
Electronic circuit assembly, 198
Electronic circuit diagram, 79
Electronic components, 127, 131
Electronic products, 97
Electronic symbols, 80
Electronic wiring, 129
Electroplating, 203, 204, 206
Elevator, 246
Emulsifier, 156
End elevation, 60
Energy, 277
Energy converter, 67
Engineer's file, 160
Engineer's try square, 140
Engineered products, 97
Engineering, 229
Engineering drawing, 21, 53
Engineering materials, 107
Engineering notation, 290

Epoxide, 250
Epoxy resin, 121, 251
Equations, 291, 317
Equilibrium, 304
Ergonomics, 16, 20
Etching, 202
Evaluation, 89
Evaluation matrix, 16, 17
Exploded view, 67, 84, 86, 230, 231, 232
Exponent, 289, 290
External client, 5
Extinguishers, 219, 220, 221
Extranet, 238
Eye protection, 213

Face-plate, 177
Fastenings, 124
Feedback, 246
Feeler gauge, 144
Ferrous metals, 113, 114
Fibre optic, 248
Fields, 235
File, 160
File card, 162
Filer, 76
Filler, 120, 208
Film resistor, 130
Filter, 75
Final control element, 245
Final design solution, 89
Fire, 218
Fire blanket, 221
Fire extinguishers, 219, 220
Firewall, 238
First aid, 224
First-angle projection, 60, 61, 62
Fitter's bench, 153
Fitter's vice, 154
Fixed steady, 180
Flexible joints, 185
Flow chart, 30
Flow control valve, 73
Flow diagram, 30
Flow soldering, 197
Fluid power schematics, 66
Fluid power symbols, 67
Fluidised bed dipping, 207
Flux, 191, 196
Flywheel, 40
Foam extinguishers, 220
Footwear, 213
Force, 304, 307
Formal drawing, 34
Four-jaw chuck, 176
Fractions, 274, 275
Frequency, 277
Fundamental units, 276

Fuse, 79
Fusion welding, 188, 189

GA drawing, 35, 36, 39
GaAs, 258
GPS, 244, 245
GRP, 121, 233, 251
Gallium arsenide, 258
Galvanising, 204, 206
Gas welding, 189
Gasket, 38
Gauge, 141, 142, 143
General arrangement drawing, 35
Generating ideas, 11
Glass reinforced plastics, 121, 251
Global Positioning System, 244
Gradient, 313
Graphs, 310, 313, 314, 315, 319
Gravity, 305
Grey cast iron, 113, 114
Grinding, 204
Grinding machines, 205
Groups of paint, 208
Guards, 210, 211
Gun metal, 117

HMOS, 254
HTML, 236, 237
HTTP, 236
Hacksaw, 154, 162
Hacksaw blade, 162, 163
Hacksaw frame, 162
Hand tap, 164
Hard soldering, 192
Hardening, 200
Hardness, 112
Hatching, 53
Hazards, 212
Headstock, 174, 178
Health and Safety at Work Act, 209
Health and safety, 106
Heat exchanger, 75, 76
Heat treatment, 200
Heatsink, 100
Hertz, 277
Hexagon, 294
High carbon steel, 114
High conductivity copper, 115
High purity aluminium, 118
Hoist, 18, 70, 71
Hole clearance, 186
Horizontal mill, 184
Hot-dip galvanizing, 206
Human engineering, 16
Hydraulic circuit, 32
Hyperlink, 236
Hypertext markup language, 236

Hypertext transfer protocol, 236
Hypotenuse, 316

I/O, 253, 261
ICT, 233
IP, 236
ISP, 236
Ideas, 11
Impact load, 111
Incentive, 7
Indices, 279, 281, 289
Inductor, 79, 132
Information and communications
    technology, 233
Input/output, 253
Inside caliper, 137
Insulators, 108
Integers, 274
Integrated circuit, 81, 134, 135, 253
Intercept, 313
Internal client, 5
Internet, 236
Internet Explorer, 236
Internet Protocol, 236
Internet Service Provider, 236
Interrupt, 262
Intranet, 238
Investigating products, 265
Investigation, 6, 15, 87
Isometric drawing, 57, 58

Joining, 184, 185
Joints, 184, 185, 194
Joule, 277

Kelvin, 276
Key, 165
Kilogram, 276

LAN, 263
LED, 248, 263
Laminated plastics, 121
Lands, 267
Landscape orientation, 25
Lap joint, 126, 194
Laser, 267
Lathe, 86, 173, 174
Laws of indices, 289
Laws of signs, 275
Layout diagram, 31
Layout drawing, 40, 43
Leader lines, 45, 46
Length, 276, 294
Letters, 46, 47
Line datum, 151, 152
Line styles, 44
Linear actuator, 70
Linear measure, 296

Lines, 42, 44, 45
Liquid plastisol dipping, 208
Local area network, 263
Low carbon steel, 114
Lubricator, 75, 76
Luminous intensity, 276

MACSS, 253
MTBF, 105
MTTF, 105
Machine guards, 210
Machine vice, 171
Magnetic properties of materials, 108
Maintainability, 105
Maintenance, 20, 211
Malleability, 111, 112
Mandrel, 175, 188
Mantissa, 289
Manufacture, 21
Manufacturing, 229, 265
Marking out, 136, 143, 146, 149
Mass, 276, 305
Materials, 106, 107, 248
Mathematics, 273
Matrix board, 128, 197, 198
Mean time before failure, 105
Mean time to failure, 105
Measurement, 136, 139
Mechanical components, 124
Mechanical engineering, 230
Mechanical products, 97
Mechanical properties of materials,
    109
Medium carbon steel, 114
Memory, 253, 262
Metals, 113
Metasearch site, 238
Metre, 276
Microlithography, 253
Micrometer, 139
Microprocessor, 253, 258, 259, 261,
    262
Microprocessor system, 261
Mild steel, 114
Milling, 183
Milling machine, 184
Mind map, 12, 13, 14
Mole, 276
Molecule, 249
Monomer, 249
Motor, 67
Mould, 183
Moulding, 121
Multiples, 276, 277
Multiplier, 277

NRV, 74, 75
Naval brass, 116

Negative indices, 281
Negative integer, 274
Netscape, 236
New technology, 233
Newsgroups, 239
Newton, 277, 305, 307
Nitro-cellulose, 122
Non-ferrous metals, 115
Non-metals, 118
Non-return valve, 74
Nonagon, 294
Notation, 279, 290
Number line, 274
Numbers, 46, 47, 273
Nut, 186, 187
Nuts, 125
Nylon, 122, 250

Oblique cutting, 155, 156
Oblique drawing, 56
Octagon, 294
Ocy-acetylene welding, 189
Odd-leg calipers, 148
Off-hand grinding machine, 205
Off-the-shelf parts, 35
Oil blueing, 206
Optical assembly, 267
Optical fibre, 248, 249
Optical receiver, 248, 249
Optical transmitter 248, 249
Optical unit, 268
Options, 11
Originator, 34
Orthogonal cutting, 155, 156
Orthographic projection, 55
Outside calipers, 137
Overheating, 201
Oxidising, 206

PCB, 31, 32, 128, 197, 198, 201,
 202
PCD, 151
PCFCV, 73, 74
PLC, 233, 263
PLC programmer, 264
PLC system, 264
PRV, 73
PTFE, 122, 207, 250
PVC, 122, 128, 208, 250
Paint system, 208
Painting, 208
Paper orientation, 24
Paper size, 24, 25
Parallel I/O, 261
Parallel turning, 178, 180
Parallelism, 180
Parts, 35, 106
Parts identification, 184

Parts integration, 252
Pascal, 277
Permanent joints, 185
Perspective, 56
Perspex, 122
Phenolic resin, 121
Phosphor bronze, 117
Photodiode, 248
Photoresist, 202
Phototransistor, 248
Pie chart, 8, 9, 10
Pigment, 208
Pillar drill, 167
Pinning, 162
Piping diagram, 32, 33
Pitch circle diameter, 151
Pits, 267
Plain washer, 187
Plan view, 60
Plane of cutting, 52
Planning, 102
Planning a drawing, 24
Plastic coating, 207
Plasticity, 110
Plastics, 250
Plastisol, 208
Plexiglass, 122
Plug, 130
Plug gauge, 142
Plumbing circuit, 33
Plys, 122
Pneumatic hoist, 70, 71
Point angle, 170
Point datum, 151, 152
Polishing, 205, 206
Polyamide, 122, 250
Polycarbonate, 267
Polyester, 250
Polyester resin, 121
Polymer, 249, 250
Polymerisation, 250
Polymers, 233
Polypropylene, 122, 250
Polystyrene, 122, 250
Polytetrafluoroethylene 122
Polythene, 122, 250
Pop riveting, 188
Portrait orientation, 25
Positive integer, 274
Power, 277
Power supply, 98, 99
Power transistor, 134
Presentation, 85
Pressure, 277, 307
Pressure compensated flow control
 valve, 73, 74
Pressure release valve, 72
Pressure relief valve, 71, 73

Primary design needs, 20
Principles, 17
Printed circuit board, 31, 198, 241
ProDesktop, 43
Product investigation, 265
Product specification, 101
Production plan, 103, 104
Production planning, 102
Products, 97
Programmable logic controllers, 233, 263
Properties of materials, 107, 108, 109
Proportion, 8
Proportionality, 286, 310
Protection of parts, 185
Protective clothing, 211, 212, 213
Protractor, 141
Putty, 208
Pythagoras' theorem, 316

Quadrant, 314, 315
Quality assurance, 105, 106
Quality control, 105
Quenching, 201
Questionnaire, 6, 7

RAM, 253, 259, 261
RCCB, 217
RCD, 214, 217
ROM, 253, 261
ROV, 255, 256, 257, 258, 260
Radians, 317, 318
Radio, 228, 229, 258, 259
Radius, 301
Radius gauge, 142, 144
Rake angle, 155
Random access memory, 253, 262
Read-only memory, 253, 262
Receders, 56
Reciprocals, 280
Records, 235
Rectangle, 297
Rectangular coordinates, 150
Reference dimensions, 50
References, 92
Refractoriness, 120
Register, 261
Reinforced concrete, 251
Reinforced plastics, 121
Relative density, 306
Reliability, 20, 105
Remotely operated vehicle, 255
Rendered 3D drawing, 40
Research, 6, 87
Residual current circuit breaker, 217
Residual current detector, 214

Resin, 121
Resistance, 82, 108
Resistor, 79, 130, 132
Resistor colour code, 132
Respirator, 213
Ribbon cable, 129
Right-angled triangle, 297
Rigidity, 112
Risk assessment, 222
Rivet length, 186
Rivet types, 127
Riveted joints, 125, 126, 187
Riveting, 111, 185, 188
Rockwell test, 112
Rounding down, 283
Rounding up, 283

SI units, 276, 278, 279
Safety education, 211
Safety valve, 71
Sand casting, 183
Saw, 162, 231
Sawing, 162
Scale, 34
Scarf and groove joint, 194
Scarf joint, 194
Schematic diagram, 31, 33, 66
Schematic drawing sketch, 26
Science, 273
Scientific principles, 17, 18
Scope of a specification, 20
Screw thread, 41
Screw thread cutting, 164
Screwdriver, 98, 99
Screwed fastenings, 124, 186
Screws, 125
Scribed line, 144, 145
Scriber, 146, 148
Scribing block, 149
Search engine, 238
Search site, 238
Second, 276
Sectional views, 52
Sectioned assembly, 53
Sections, 52
Self-locking nut, 187
Sensor, 248, 263
Second, 276
Septagon, 294
Sequence valve, 73
Serial I/O, 261
Servicing, 20
Set, 163
Set point, 245
Shank, 168
Shapes, 294
Shock hazard, 215, 216
Signal, 247

Significant figures, 283
Silicone, 250
Silver solder, 192
Silver steel, 115
Sine, 317, 319
Sketch, 27
Sketching, 26
Slip gauge, 149
Smart materials, 233
Socket, 130
Soft soldering, 190
Solar panel, 258
Soldered joint, 191, 192, 196
Soldering, 190, 191, 192, 198
Soldering flux, 191
Solid-state device, 134
Solution, 16
Solutions, 15
Solvent, 208
Spanner, 165
Specification, 19, 88, 100, 101
Speed, 308
Sphere, 302
Splice joint, 126
Spot facing, 172, 173
Spring washer, 187
Square, 294, 297
Square law graphs, 313
Square roots, 281
Standard form, 289
Standards, 41
Steel rule, 136, 137
Straight grinder, 205
Straight line graphs, 310, 312
Straight line law, 313
Straight shank, 168
Strainer, 75
Strength, 109
Strip board, 128, 197, 198
Sub-multiples, 276, 277
Substrate, 267
Surface coating, 204
Surface datum, 151, 152
Surface finishing, 204
Surface plate, 149
Surface table, 149
Surfacing, 178, 181
Survey, 6, 7, 9
Swarf, 171
Sweating, 192
Switches, 81
Symbols, 276
Systems, 245

Tab washer, 187
Tailstock, 86, 174, 175, 182
Tang, 169
Tangent, 317, 319

Tap, 164
Tap wrench, 164
Taper mandrel, 175
Taper shank, 168
Taper washer, 187
Tapping, 181, 182
Tapping size, 164
Target group, 6
Technical report, 91, 92
Teflon, 207, 250
Telecommunications, 230
Telemetry, 258
Temperature, 276
Tempering, 201, 202
Template, 34
Thermal properties of materials,
     109
Thermistor, 132, 134
Thermoplastics, 121, 250
Thermosets, 120, 250
Thermosetting plastics, 120, 121,
     250
Thinner, 208
Third-angle projection, 60, 63, 64,
     65
Thread, 41
Threading, 182
Three-jaw chuck, 176, 177
Tin bronze, 117
Tinning, 190
Tolerance, 50
Tolerances, 142, 184
Tolerancing, 141
Tongue and groove joint, 194
Top coat, 208
Tough pitch copper, 115
Toughness, 110
Track layout, 31, 32
Transceiver, 232
Transducer, 245, 248
Transformer, 80
Transistor, 134, 135
Transistor amplifier, 83
Transistors, 81
Travelling steady, 180
Triangle, 294, 298
Triangulation, 244
Trigonometrical equations, 317
Trigonometrical functions, 319
Trigonometrical ratio, 317
Trigonometry, 316
Try square, 140, 148
Tufnol, 121, 202
Turning, 173, 178, 180
Turning operations, 178
Turning tools, 178, 179
Twist drill, 169, 170
Two-way lighting switch, 82

UHF, 259
URL, 236
Under heating, 200
Undercoat, 208
Uniform resource locator, 236
Unit conversion, 278
Units, 276
Urea formaldehyde, 250

VLSI, 254, 261, 262
Valve, 67, 72, 73, 74
Variables, 285
Vehicle, 208
Velocity, 277, 308
Vernier calipers, 138
Vertical mills, 184
Vice, 153, 154, 171
Vickers test, 112
Vinyl plastics, 122
Voltmeter, 81
Volume, 302

Washer, 187
Water trap, 75, 76

Watt, 277
Web address, 236
Web browser, 236, 237, 238
Web directory, 238
Web page, 236, 237
Wedge angle, 154
Weight, 305
Weld zone, 190
Welding, 188, 189
Wire frame drawing, 242
Wirewound resistor, 130
Wiring, 129
Wiring diagram, 31
Witness marks, 147
Work holding, 171, 172, 174
World Wide Web, 234, 236
Wrench, 164

Yield strength, 109

Zinc, 115, 117